英国鲁班中餐烹饪艺术

天津市经济贸易学校　编著
（天津市烹饪技术学校）

中国商业出版社

图书在版编目（CIP）数据

英国鲁班中餐烹饪艺术/天津市经济贸易学校（天津市烹饪技术学校）编著．——北京：中国商业出版社，2021.11
 ISBN 978－7－5208－1510－9

Ⅰ.①英… Ⅱ.①天… Ⅲ.①中式菜肴－烹饪 Ⅳ.①TS972.117

中国版本图书馆CIP数据核字（2020）第259783号

责任编辑：李 飞 蔡 凯

中国商业出版社出版发行
010－63180647 www.c—cbook.com
（100053 北京广安门内报国寺1号）
新华书店经销
北京广达印刷有限公司印刷

*

889毫米×1194毫米 16开 22印张 400千字
2021年11月第1版 2021年11月第1次印刷
定价：138.00元

* * *

（如有印装质量问题可更换）

编写说明

《英国鲁班中餐烹饪艺术》一书由天津市经济贸易学校（天津市烹饪技术学校）中餐烹饪专业教学团队和英国超卓教育顾问服务有限公司专家团队共同编写完成。该教材涵盖中国饮食礼仪文化、盘饰及冷菜制作、热菜制作、面点制作、津派面塑相关内容的介绍，以支持英国鲁班工坊中餐烹饪艺术专业二级、三级课程内容的学习。

教材依托学校开发的中餐烹饪教学标准、中式烹调师培训内容，突出"以职业活动为导向、以职业能力为核心"的指导思想，按照"理实一体"、"做中学、做中教"的职业教育教学理念编写。涵盖中国饮食文化的发展，即中式烹调工艺、热菜制作、烹饪原料基础知识、烹饪原料加工技术、冷菜制作工艺、冷菜盘式及装饰工艺、津派面塑制作工艺、礼仪服务及茶文化等基础知识和基本技能。以"实际、实用、实践"为原则，由浅入深，突出对实践技能的培养。通过任务教学法激发学习者的学习兴趣和创造力，使其掌握一定的专业知识、专业技能，以详实、精美的菜例激发学习者对理论知识与专业技能的活学活用，以满足职业岗位能力培养的需要。

教材在编写过程中，以培养国际化中餐烹饪技能型人才为目的，以应用优先、广度优先、利学优先为基础，突出技能训练、能力培养，满足了学习者多形式、多角度的学习和练习要求，对中餐烹饪基础技能掌握具有极强的指导作用。

因编写仓促，水平有限，书中难免有错误和不足，希望广大读者不吝赐教。

Contents

Forward 1

Introduction to Lu Ban Workshop Project in the UK 2

Part One: Chinese Food Culture 1

  Chapter 1 The Evolution of Chinese Food Culture 1

    Section 1 The Budding Period 1

      1.1 The Development of Food Culture During the Budding Period 2

      1.2 The Characteristics of the Budding Period 5

    Section 2 The Formation Period 5

      2.1 Further Development of Agriculture Broadened Food Sources 5

      2.2 Specialised Craftsmanship Allowed Culinary Tools to Upgrade Constantly 6

      2.3 Innovation in Culinary Techniques Led to the Creation of a Greater Variety of Cuisine ... 6

      2.4 Initial Form of Banquets and Food Market Started to Emerge 6

    Section 3 The Development Period 6

      3.1 Food Sources Continued to Increase 6

      3.2 New Breakthroughs in Energy Sources, Cooking Utensils, and Tableware 7

      3.3 Constant Innovation in Culinary Skills 9

      3.4 The Popularity of Banquets and Food Markets 10

      3.5 The Improvement in Chinese Dietary Theory 10

    Section 4 The Maturation Period 11

      4.1 Extensive Food Sources 11

      4.2 Exquisite Tableware 11

      4.3 A Relatively Complete System of Culinary Skills 12

      4.4 The Emergence of Regional Cuisines 12

      4.5 The Establishment of a Complete Banquet Culture 12

    Section 5 The Contemporary Period 13

      5.1 Advanced Kitchen Appliances 13

5.2 Industrialisation of Food Production 13

5.3 Food Sources Continue to Grow 14

5.4 Exchange of Culinary Resources Nationally and Globally 14

5.5 The Trend of Chinese Food Development 14

Chapter 2 The Characteristics of Chinese Food Culture 14

Section 1 Characteristics of Chinese Food Culture 14

1.1 A Wide Range of Cooking Ingredients and Superb Culinary Skills 14

1.2 A Wide Range of Cuisine Styles 16

1.3 A Manifestation of Chinese Culture 17

1.4 Combined Effect of Food and Medicine for Yangsheng (Nourishing Life) 17

Section 2 The Guiding Ideology of Chinese Traditional Diet 18

2.1 Contents and Embodiment 18

2.2 The Development of Food Science in Modern China 20

Section 3 Traditional Chinese Diet Composition 20

3.1 Traditional Chinese Diet Composition 20

3.2 The Validation and Limitation of Chinese Traditional Diet Composition 21

Chapter 3 The Classification of Chinese Cuisine 21

Section 1 The Classification of Chinese Cuisine from a Historical Perspective 21

1.1 Folk Cuisine 22

1.2 Imperial Court Cuisine 25

1.3 Scholar-officials Family Cuisine 27

1.4 Temple Cuisine 30

1.5 Ethnic Cuisine 32

1.6 Marketplace Cuisine 33

Section 2 Schools of Chinese Cuisine 34

2.1 Sichuan Cuisine 34

2.2 Shandong Cuisine 37

2.3 Jiangsu Cuisine 42

2.4 Guangdong Cuisine 47

2.5 Beijing Cuisine 53

Chapter 4 Chinese Food Related Folk Customs and Etiquettes 57

Section 1 Daily Food Customs 57

1.1 Key Features of Chinese Daily Food Customs 57

1.2 Daily Food Customs of Han Chinese 57

Section 2 Food Related Folk Customs in Festivals 61

Contents

    2.1 Chun (Spring) Festival 61

    2.2 Qingming (Tomb-sweeping Day) Festival 63

    2.3 Duanwu (Dragon Boat Festival) Festival 65

    2.4 Zhongqiu (Mid-Autumn) Festival 66

    2.5 Chongyang (Double Ninth) Festival 68

    2.6 Dongzhi (Winter Solstice) Festival 70

    2.7 La Ba Festival 72

  Section 3 Food Customs in Life's Occasions 72

    3.1 Celebration of Birth 73

    3.2 Traditional Chinese Wedding Celebration 73

    3.3 Celebration of Longevity Party 75

    3.4 Funerary 75

Chapter 5 Customs and Etiquette in Chinese Dining 76

  Section 1 Common Forms of Chinese Dinners 76

    1.1 Banquet 76

    1.2 Informal Meals 77

    1.3 Business Meals 77

    1.4 Buffet 77

  Section 2 Chinese Banquet Arrangements 78

    2.1 Definition 78

    2.2 The Common Banquet Arrangements 78

    2.3 Common Dish Combination in Chinese Banquet 79

  Section 3 Chinese Banquet Etiquette 83

    3.1 Seating Arrangement and Table Layout of Chinese Banquet 83

    3.2 The Placement and Use of Tableware 84

    3.3 Dining Etiquette in a Chinese Banquet 90

  Section 4 Tea and Wine Culture in Chinese Banquets 91

    4.1 Wine Drinking Etiquette in Chinese Banquets 91

    4.2 Tea Etiquette in Chinese Banquets 93

Chapter 6 Outline of the Regional Culture, Food History & Culinary Style, Tianjin 94

  Section 1 An Overview of Tianjin Food Culture 94

  Section 2 Tianjin Local Food Culture 94

  Section 3 Tianjin Local Cuisine 95

    3.1 The Formation and Development of Tianjin Local Cuisine 95

    3.2 Characteristics of Tianjin Cuisine 96

        3.2.1 Specialised in Cooking River Fish and Shrimp and Seafood Ingredients 96

3.2.2 Eating According to Season 96

3.2.3 Excellent Seasoning Techniques 96

3.2.4 Unique Cooking Techniques 97

3.2.5 Great Variety 97

Section 4 Representative Dishes of Tianjin Cuisine 97

4.1 The Best Restaurants and Their Notable Dishes 97

4.2 Famous Tianjin Street Food Food... 101

Part Two: Ingredients & Equipment.. 107

Chapter 7 Ingredients & Equipment 107

Section 1 Commonly used ingredients in Chinese Cooking 107

Section 2 Large Kitchen Equipment 116

2.1 Deep Fat Fryers 116

2.2 Atmosphere Steamers 116

2.3 Bain Marie.. 116

2.4 Bratt Pans 117

2.5 Griddle 117

2.6 Blender 118

2.7 Mixer 118

2.8 Processor 118

2.9 Conventional Ovens 118

2.10 Combination Stoves 118

2.11 Salamander.. 119

2.12 Industrial Wok Burner Chinese Stove 119

2.13 Industrial Induction Chinese Wok Cooker 119

2.14 Rice Cookers 119

Section 3 Chinese Kitchen Cooking Utensils 120

3.1 Wok 120

3.2 Chinese Cooking Ladle 120

3.3 Chinese Scoop Strainer 121

3.4 Kitchen Sieves 121

3.5 Wok Shovel 121

3.6 Oil Drum... 121

3.7 Metal of Bamboo Steamer 121

3.8 Cooking Chopsticks 121

3.9 Wok Brush... 122

    3.10 Chinese Rolling Pin 122

Part Three: Chinese Culinary Techniques 123

Chapter 8 Fundamental Knife Skills and Cutting Techniques 123

Section 1 Introducing Chinese Knives (by function) 123

    1.1 Slicing Knife 124

    1.2 Cutting Knife 124

    1.3 Cleaver (Chopping Knife) 124

    1.4 Cutting-Chopping Knife 124

    1.5 Roast-Duck Knife (Small Slicing Knife) 124

    1.6 Kitchen Scissors 124

    1.7 Carving Knives 124

Section 2 The importance of Cutting Techniques in Chinese Cooking 125

    2.1 Functions of Cutting Techniques in Chinese Cooking 125

    2.2 Requirements of Applying Cutting Techniques 125

Section 3 Cutting Techniques 127

    3.1 Straight Cutting Technique 127

    3.2 Flat Slicing Technique 131

    3.3 Oblique Slicing Technique 132

    3.4 Scoring Technique 132

    3.5 Other Cutting Techniques 133

Section 4 Knife Skills 135

    4.1 Chunks 135

    4.2 Slices 136

    4.3 Shreds 136

    4.4 Strips 137

    4.5 Dice, Fine Dice, Mince 137

    4.6 Mash and Paste 137

    4.7 Patterns 138

Chapter 9 Chinese Cooking Techniques 138

Section 1 Chinese Hot Dish Cooking Techniques 138

    1.1 Zha (Deep Frying) Techniques 140

    1.2 Chao (Stir-frying) Techniques 144

    1.3 Liu 148

    1.4 Bao 150

    1.5 Jian (Pan Frying) 152

  1.6 Ta 153

  1.7 Peng 154

  1.8 Shāo 155

  1.9 Mènn 158

  1.10 Kao 160

  1.11 Dùn 160

  1.12 Zheng 162

  1.13 Hui 164

  1.14 Cuan 165

  1.15 Zhu 166

  1.16 Shuan 168

  1.17 Basi (Caramelising/Candied) 169

  1.18 Ju (Baking) 170

 Section 2 Chinese Cold Dish Cooking Techniques 171

  2.1 Ban (Tossing) 171

  2.2 Qiang 172

  2.3 Yan (Marinating) 172

  2.4 Pao (Pickling) 173

  2.5 Boiled in plain water 174

  2.6 Boiling in salted water 174

  2.7 Lu 175

  2.8 Jiang 176

  2.9 Dong (Aspic) 176

  2.10 Frying – Lu 177

  2.11 Hot Qiang 178

  2.12 Liu Li (Caramelising) 179

  2.13 Zao 179

 Section 3 Chinese Pastry Making Techniques 179

  3.1 Flavours and Styles of Chinese Pastry 180

  3.2 Dough Making Techniques 181

 Section 4 Modern Cooking Methods 183

 Section 5 Chinese Culinary Vocabulary 184

 Section 6 Allergen 187

Chapter 10 Recipes 188

 Section 1 Fish & Shellfish Recipes 188

 Section 2 Meat & Poultry Recipes 212

Section 3 Vegetable Recipes	249
Section 4 Noodles, Dumplings, and Flatbread Recipes	256
Section 5 Chinese Soup Recipes	301
Section 6 Chinese Dessert Recipes	312
Section 7 Plate Presentation Ideas	338
Chapter 11 Dough Modelling	342
Section 1 An introduction to Tianjin Style Dough Modelling	342
Section 2 Materials & Tools for Dough Modelling	345
Section 3 Fundamental Dough Figurines Making Techniques	346
3.1 Carving and Shaping Methods	346
3.2 Fundamental Modelling Methods	348
Conversion tables	349

Forward

The ***Lu Ban Chinese Culinary Arts*** is the first of its kind in many ways. The book looks way beyond the European perception of Chinese cooking and dinning and goes much deeper into the true art of Chinese culinary arts, history, culture, and philosophy.

The book takes you on a culinary journey starting over two thousand years ago and guides you through the intriguing history of Chinese dynasties, landscape, cultural differences, and its people. This jigsaw of complex folk customs, religion, and ideology alongside a rich abundance of locally sourced foods has made Chinese cuisine one of the most complex and varied cuisines in the world today.

The book microscopically explores this journey of discovery and evolution of food within China and pays special addition to the region of Tianjin. Explaining how a country with a population of 1.433 billion people has managed to feed its people independently, without any outside influence or support for over two thousand years. One discovers that Chinese food to its people is much more than just sustenance, food is the epicentre of Chinese life and wellbeing. What you eat, when you eat, how you eat and who you eat with, all have significant importance within Chinese culture and social order.

Alongwith this impressive and interesting food culture education, the book offers the readers a repertoire of truly authentic Chinese dishes, each one written and developed by Chinese master chefs and carefully translated and tested by leading UK chefs. Bringing carefully crafted and hand picks recipes to the book.

The book has been purposely written in a style to support the professional qualifications that have been designed to run alongside it. The Chinese Culinary arts programs at both levels 2 and levels 3 are the first qualifications of their kind outside of China. Their overarching aim is to teach authentic Chinese culinary arts to the next generation and to help honour and respect the true authenticity of Chinese cuisine.

As both a chef and an educator for over thirty-five years, I believe in the importance of good food and understand the many intrinsic and extrinsic factories related to food and its consumption. For me what is so unique about this book is that it looks so far back into why food is so important. And by doing this, it shines a light on the way forward and maybe reminds us all how to get the very best out of our dining experiences.

Introduction to Lu Ban Workshop Project in the UK

Lu Ban Workshop

Lu Ban Workshop project was initiated by the Tianjin Municipal Education Commission in 2015 aiming to share Chinese technical and vocational techniques and skills with the world. The project was named after the Greatest Craftsman in Chinese history, Lu Ban, an inventor, carpenter, and engineer who demonstrated strong expertise, creativity, relentless pursue of excellence, inspiring generations of artisans to follow.

Lu Ban Workshop project was introduced to the UK in 2017 by PAM Education with the mission to enable education and training in authentic award winning Chinese culinary skills in the UK, to improve the quality of Chinese food and its preparation worldwide, to promote Chinese food culture, and to create high-end brands of Chinese cuisine. In addition, we hope to build a network across food industry to allow better cooperation between the business sector and the vocational education sector both in China and in the UK.

The success of Lu Ban Workshop project in the UK was credited to a multilateral partnership including education and business partners from both countries. A set of UK fully regulated qualifications in Chinese culinary arts and their associated learning and teaching resources in English have developed.

Timeline

2016
- December 2016, PAM Education officially established its strategic partnership with Tianjin Economics and Trade School (then the Tianjin No. 2 School of Commerce & Tianjin School of Cuisine) to bring authentic Chinese culinary arts training and education onto the global stage.
PAM Education Consultancy Services Ltd. is an independent education solution provider with the main objective of bridging the gap between vocational education and industry development and employability skills.

2017
- The partnership has successfully obtained funding from Tianjin Municipal Education Commission under the Lu Ban Workshop Project to establish education and training centre in the UK for Chinese Culinary Arts, to develop fully regulated qualifications and the associated learning resources in English.
- In May, the first Lu Ban Workshop in Europe was opened at Chichester College (CC). Officials from the Tianjin Education Commission, Senior management from PAM, TES, and CC were participated in the opening ceremony.

2018
- In April, the first and only UK fully regulated Level 3 Diploma in Chinese Culinary was approved by PAM Education's awarding Body - Qualifi and appeared on the Regulated Qualification Framework.
- In October, PAM Education offered 10 scholarships to prototype the L3 Diploma course with Chichester College Group. Most learners reported to have benefited from the course professionally.

2019
- In Early January, PAM-TES collaboration continued to flourish. Witnessed by the senior management team of the Tianjin Food Group, PAM Education and TES signed a MoU to develop a restaurant and international training centre at the iconic building of Cains Brewery in the heart of Liverpool's vibrant Baltic Triangle.
- On 31st January, PAM's Chinese Culinary Arts students were given the opportunity of a lifetime when they cooked for Number 10 Downing Street and the Rt Hon Theresa May at the 2019 Chinese New Year Reception. With 150 people attending the event, the students excelled in cooking a variety of canapés alongside four prestigious Chinese master chefs, flown in to help prepare for the event.
- In July, the Level 2 Certificate in Chinese Culinary Art was approved.
- In October, the Level 4 Diploma in Chinese Culinary Management was approved.
- In November, Lu Ban Restaurant and Training Centre Liverpool was launched, which have not only brought premium Chinese cuisine and dining experience to Liverpool but also provide the state of art learning venue and work placement opportunities to Chinese culinary lovers.
- BBC North West and the BBC the One Show reported on the ground-breaking Chinese culinary experience

that the Lu Ban restaurant and training centre will provide.

2020
- In July, Lu Ban Workshop project partners initiated the Chinese Master Chef Apprentice Training Programme to upskill UK chefs.

Part One: Chinese Food Culture

Chapter 1 The Evolution of Chinese Food Culture

Humans gradually changed their diet from eating raw food to cooked food when they discovered the benefits of cooking with fire. Cooking enabled humans to transform natural edible objects into cuisine. Distinctive food culture started to form as a result. In China, the history of food culture is regarded to have gone through the Budding Period, Formation Period, Development Period, and Maturation Period.

Section 1 The Budding Period

In pre-historic times, humans survived on natural edible objects such as fruits, roots, leaves, animals including fish, birds and insects. It is commonly described as a time of eating raw meat and drinking blood. After a period of many centuries, mankind discovered that meat of wild animals and nuts were easier to chew and tasted better after being cooked by fire. Gradually early, humans learned the ways to create fire by friction. The ability to use and control fire helped early humans to understand that food could be cooked for different periods of time

Part One

and in a variety of ways.

China has a long tradition involving food culture in Chinese Methodology. Legendry stories often illustrate the tradition of food. It is believed that Chinese food culture started to shape during the Neolithic Age from 6000 BC to 2000 BC.

The legendary ruler "Yellow Emperor" (2717BC-2599 BC) is regarded as the initiator of Chinese civilization and is

said to be the ancestor of all Chinese. He is arguably the earliest form of Kitchen God to be worshiped by Chinese people. Among his numerous inventions and innovations, he was credited with the invention of the earliest steamer and the way to make a stove to intensify the heat. As a result, fuel was saved, and the food was cooked thoroughly in less time.

During the period of the Yellow Emperor, there was another important legendary figure called Su Sha who was a smart and capable hunter. People associated him with the method of extralting salt from sea water. The story says that one day whilst he was cooking fish in a pot using sea water he saw a wild boar and decided to chase and kill it. Upon his return, he found the sea water in the pot had boiled away, leaving a layer of salt. He tasted it and the taste was amazing. He then dipped a piece of grilled wild boar meat in the salt to find the meat was much tastier. It is said that Chinese people have since used salt with food to add flavour and for it to be healthier.

1.1 The Development of Food Culture During the Budding Period

During the long budding period, Chinese food culture development can be observed in the following ways:

Reliable Food Sources

The origins of rice and millet farming date to the same Neolithic period in China. Ancient Chinese people pioneered the cultivation of a range of staple crops which were referred to as the "Five Grains" or "Hundred Grains" including millet, broomcorn millet, wheat, bean, sesame, and rice.

Agriculture triggered such a change in society and the way in which people lived that its development has been dubbed the "Neolithic Revolution". The world's oldest known rice paddy fields, discovered in eastern China in 2007, reveal evidence of ancient cultivation techniques such as flood and fire control. Farming undoubtedly had greatly affected the evolution of Chinese food by providing reliable food supply.

Chinese Food Culture

The Invention of Cooking Stove

The earliest clay stove in China appeared during the Middle to Late Neolithic Age (about 5000 BC to 1500 BC). It was cast in clay and fired in kilns the same way as pottery vessels' by Yangshao Culture (ca. 5000 BC to 3000 BC), one of the earliest settled cultures in China. The clay stove was portable. "Fire stove", an enhanced version of the early stove, was introduced to people by the Yellow Emperor (2717 BC - 2599 BC) to intensify the heat. Fire stoves were built on the ground.

The Invention of Pottery Vessels

The earliest pottery vessels date back to 20000 BC and were discovered in Xianrendong cave in Jiangxi, China. The pottery may have been used as cookware. Other early pottery vessels include those excavated from the Yuchanyan Cave in southern China, dated from 16000 BC.

By the Middle and Late Neolithic, most of the larger archaeological cultures in China were settled farmers, who produced a variety of attractive and often large vessels. Many of them are believed to have been used as cook-

Part One

ware. There were four types of pottery cookware based on their functionality in cooking.

1. Water vessels: such as GUAN (pot), FU (similar to cauldron) and DING (similar to casserole with three or four legs), these were used for boiling water, making porridge and soup as well as meat casseroles.

2. Steam vessel: The Chinese name of the ancient steam cooker is Zeng. It is a basin with holes on the bottom. It was usually placed on Fu or Ding to steam food.

3. Baking vessel: the design of the baking vessel allowed heat to bake the food.

4. Grilling vessel: by using this type of vessel, food can be grilled in much the same way as modern equipment today.
5. The invention of pottery and the emergence of massive pottery production has revolutionized Chinese food culture.

The Use of Flavourings

The use of flavorings was a follow-on to the invention of pottery vessels. Flavorings were used and developed to enhance taste and remove fishy and meat odours. The method of seasoning has greatly enriched the perception of taste. Hence, the use of salt was considered the second most important breakthrough in dietary history after the use of fire from a historical perspective.

1.2 The Characteristics of the Budding Period

The "Budding period" of Chinese Food Culture lasted over 4,000 years during which time early humans made significant developments in the use and control of fire and cultivation, preparation, as well as consumption of food.

Table 1 Summary of the Development of Chinese Food Culture in the Budding Period

Cultural Aspects of Food	The Characteristics of the Budding Period
Use and control of fire for cooking	Early humans discovered fire and gradually learned to make fire by friction, using fire for cooking food; control fire to cook food for different periods of time. The cooking stoves were invented to intensify the heat
Cultivation of food	Primitive farming gradually replaced hunting, fishing, and gathering wild plants to become the primary food source which provided humans with a reliable supply of materials for cooking
Preparation of food	Pottery utensils were created and used for cooking and storage. Basic cooking methods such as steaming and boiling were invented accompanied by some simple cooking skills. The concept of seasoning by adding different flavourings into food started to develop
Consumption of food	Clay pottery utensils might have been used for eating

Section 2 The Formation Period

The Formation period of Chinese Food Culture between the Xia Dynasty (2070 BC - 1600 BC) to the Spring and Autumn period (771 BC - 476 BC) and the Warring states period (770 BC - 221 BC) lasted around 2000 years and coincided with significant changes in China's political, economic, and cultural life. The Formation Period witnessed the Chinese food culture to evolve in the following four ways.

2.1 Further Development of Agriculture Broadened Food Sources

During the Xia Dynasty, the agricultural economy started to emerge in China. It was largely advanced in the Shang Dynasty (1646 BC - 1600 BC) as a result of religious need to supply many thousands of buffalo, sheep, and pigs for animal sacrifice.

During the Spring and Autumn period and the Warring states period, cast iron tools and ploughs pulled by buffalos were widely used. The large scale harnessing of rivers and development of water conservation projects were developed during this period. Animal farming was relatively advanced at the time. Archeological discoveries evidenced the existence of fish farming and poultry raising.

Part One

All in all, the advances in agriculture provided people with a wide variety of food.

2.2 Specialised Craftsmanship Allowed Culinary Tools to Upgrade Constantly

Over the course of the Formation period, tool making craftsmanship reached a new height. In the Xia Dynasty, for example, bronze utensils for cooking food with oil were crafted. Cooks' knife skills were far superior owing to the invention of bronze knives.

In the Shang Dynasty lacquerware was introduced and by the time of the Warring States Period, the lacquerware became quite exquisite, together with ivory ware, bone ware and jade ware often appeared in banquets and sacrificial rites.

2.3 Innovation in Culinary Techniques Led to the Creation of a Greater Variety of Cuisine

With the improvement of wet based cookery methods, such as boiling and steaming and the introduction of techniques with dry-heat such as grilling, and baking cookery improved. The addition of deep frying and the ability to thicken with cornstarch, increased variety, and innovation.

2.4 Initial Form of Banquets and Food Market Started to Emerge

Banquets were started as a result of people taking part in the celebration of ancestors and religious ritual. Increased productivity resulted in a greater supply of products which in turn promoted the opportunity for trading in the emerging markets. According to the written records, a roadside hut was built to serve as the place for trading goods every 5,000 meters and every hut provided food.

Section 3 The Development Period

The period from the Qin Dynasty (221 BC -207 BC) to the Tang Dynasty (907 BC- 618 AD) is regarded as the development period of Chinese food. In the 1,200 years, Chinese people made significant development in all aspects of food. This period had left behind many remarkable culture heritages and paved the way for Chinese food to reach its maturity.

3.1 Food Sources Continued to Increase

After the State of Qin unified ancient China, productive of the country was improved a great deal. During this period, food came from sources including agriculture, fish, meat and wild plants gathering.

New food was also introduced in substantial amounts from outside China. Especially followed by Zhang Qian's imperial expedition to Central Asia in the 2nd century BC during the time of the Han Dynasty (202 BC - 220 AD), food such as cucumbers, lima beans, peppers and spinach were brought in. These sources provided new ingredients for Chinese cuisine to thrive.

Tofu-making was first recorded during the Han Dynasty. The invention of tofu is a notable contribution to both Chinese and global food. Before the Han Dynasty, people cooked with animal oils. Since the Han Dynasty, vegetable oils have been more widely used.

In terms of seasoning, fermented soya beans and caramel were newly added. During the Wei, Jin, and Southern and Northern Dynasties (222 AD - 589 AD), orange peel, ginger, spring onions, garlic, peppers were all used for seasoning. As for seafood, sea crabs, flatfish, and jellyfish were listed in the recipes during the Tang Dynasty.

3.2 New Breakthroughs in Energy Sources, Cooking Utensils, and Tableware

The Breakthrough in Energy Sources

In the period from the Han Dynasty to the Tang Dynasty, the new breakthrough in energy sources lay in the use of coal. China was the first country in the world to use coal as a fuel. The practice of cooking with coal can be traced back to the Eastern Han Dynasty (25 AD - 220 AD) but this practice was not widespread until the Northern and Southern Dynasties and only by the time of the Tang Dynasty, was coal commonly used.

The Breakthrough in Cooking Utensils

During the Han Dynasty, Chinese ironworking achieved high levels of sophistication. Iron was the material of choice throughout China for most tools and weapons. Iron pans and knives provided the foundation for the development of Chinese cooking methods and knife skills.

Part One

The Breakthrough in Tableware

During the Tang Dynasty, there was a large amount of ceramic tableware in use, utensils were clean and eye-catching, adding colours to dishes. The five Chinese culinary aesthetic aspects to appreciate cuisine, i.e. colour, aroma, taste, presentation, and tableware contributed to the food culture during this period.

3.3 Constant Innovation in Culinary Skills

After the Qin and Han Dynasties, culinary skills were categorised into knife skills and cooking techniques. Dishes and pastry were distinguished. The specialisation resulted in a dramatic improvement in culinary skills. For example, it was said that the meat can be cut as thin as a Cicada's wings (butterfly's) to reveal the fibers inside. During the Sui and Tang Dynasties (581 AD - 907 AD), a variety of new dishes were produced. The Seven Wonders of Jiankang represented the advanced pastry-making skills and It was recorded in the **Qingyilu** (Records of the Unworldly and the Strange) written by Tao Gu (950 AD).

Table 2 New Culinary skills Innovation in the Development Period

Dynasty	Cooking methods Invented			Description
	In Chinese	In Pinyin	In English	
The Han Dynasty	杂烩	Zahui	Stew	A rich soup or stew containing meat, seafood and vegetables
	涮	Shuan	Blanch	To cook by dipping finely sliced ingredients briefly in boiling that is or soup (generally done at the dining table) a dish where thinly sliced meat and vegetables are boiled briefly in a broth and then served with dipping sauces
The Tang Dynasty	冷淘	Lengtao	Refresh	Boiled noodles to be washed in iced water, seasoned and kept at a low temperature or on ice before consumption. A typical and popular type of noodle dish in summer
	冰制	Bingzhi	Frappe, sorbet	Iced snacks
The Southern and Northern Dynasty	糟	Zao	Pickled	Preserved in vinegar or alcohol with seasonings
	酱	Jiang	Slow red	Also known as Chinese stewing, red braising, red stewing. This technique imparts red colour with the addition of either, red fermented tofu, fermented bean paste, soy sauce or caramelised sugar
	炒	Chao	Stir-fry	Cooking small pieces of even sized ingredients at high temperature with the addition of oil or fats

Figure 1 Su Shan was served at a banquet in the Tang Dynasty.

Part One

3.4 The Popularity of Banquets and Food Markets

During the Development Period, staging a feast became increasingly popular, a tradition followed by the noble court of the Emperors and common people. Particularly during the Sui Dynasty and the Tang Dynasty, exquisite dishes were served at banquets in locations with scenic views and or hosted on boats. Venues along the Qujiang River in Chang'an, the capital of the Tang Dynasty, were the most popular.

The growth of food markets mirrored the flourishing economic and cultural development during a period from the Qin and Han Dynasties to the Sui and Tang Dynasties. The great development of agriculture and handcrafts coincided with the expansion of towns and cities. In this boom period for business, a thriving hotel and restaurant industry evolved. Food markets opened all night long in cities such as Chang'an, Yangzhou, Suzhou and Hangzhou.

3.5 The Improvement in Chinese Dietary Theory

During the Development Period, staging a feast became increasingly popular, a tradition followed by the noble court of the Emperors and common people. Particularly during the Sui Dynasty and the Tang Dynasty, exquisite dishes were served at banquets in locations with scenic views or hosted on boats. Venues along the Qujiang River in Chang'an, the capital of Tang Dynasty, were the most popular.

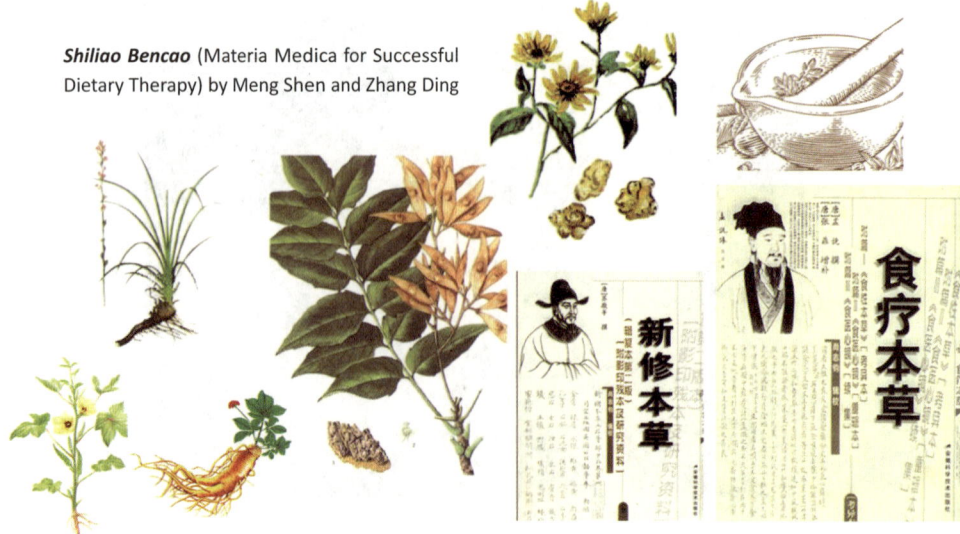

Shiliao Bencao (Materia Medica for Successful Dietary Therapy) by Meng Shen and Zhang Ding

Another notable food writing was **Shiliao Bencao** (Materia Medica for Successful Dietary Therapy) by Meng Shen and Zhang Ding.

Kanhuai Fa, the first book that is dedicated to knife skills, was also written in the Dang Dynasty.

Section 4 The Maturation Period

Significant changes had taken place in China during the period of the Song (960 AD-1279 AD), the Yuan (1206 AD-1368 AD), the Ming (1368AD-1644AD), and the Qing (1636 AD-1912 AD) dynasties from economic, political, and cultural perspectives. Accompanied by these changes, Chinese Food Culture stepped into the maturation period. The characteristics of this period can be summarized from the following 5 aspects.

4.1 Extensive Food Sources

The list of food sources is shown below:
- Newly discovered species of wild animals and plants
- New breeds created through innovative cultivation
- Introduction of sweet potatoes, tomatoes, pumpkins, kidney beans, potatoes, and cauliflower food from other countries
- Seafood became a popular food choice as a result of advanced navigation and water transportation during the Yuan Dynasty (1206 AD-1368 AD). Shark fins and sea cucumbers were considered as the high-value cooking ingredients at banquets.

4.2 Exquisite Tableware

Multi-layered steamers were invented to save the cooking time in the Song Dynasty.

The Yuan, the Ming, and the Qing Dynasties represent an extraordinary period of innovation in ceramic production in Chinese history. Porcelain became a booming industry with Jingdezhen (the town of Jingde) developed into the most well-known centre for producing the highest quality porcelain in China.

Cookware and tableware were designed to enhance their functionality and aesthetics.

Exquisitely decorated gold and silver tableware were produced in large amount with improved quality. For example, in the Qing Dynasty, the wine pot exclusively used by royalty was embellished with a variety of embossed decorative patterns of flying dragons and auspicious clouds to add glamour and splendor to the wine pot.

4.3 A Relatively Complete System of Culinary Skills

During the Song and the Yuan Dynasties, there were new cooking techniques developed.

Pastry Dough Making Methods
- Leavened dough and puff pastry dough were also achieved.
- Cold, hot and boiling water were used in dough making to attain various levels of consistency.

Noodle Making Methods and Skills
- It was recorded in **Sushi Shuolue** (Vegetarian Dietary Impression) by Xue Baocheng in the late Qing Dynasty that Chenmian (捫面) noodles were made in a range of shapes including extruded triangle, cylinder, fine threads etc.

Food Cutting Methods and Skills
- Food was presented in artistic style. Flower shapes were used in food carving.
- The knife skill involved deboning a chicken without changing its bodily form was created during the Ming Dynasty.
- Melons were cut into a bowl shapes to contain food during the Qing Dynasty. This shape had a unique Chinese name called Zhong (盅).

Seasoning Methods and Skills
- During the Yuan Dynasty, red yeast rice was used to colour a wide variety of food products.
- Zhaoyou sauce (糟油), fermented bean curd, Sharen (砂仁，Amomum Fruit), and Sichuan Pepper (花椒) were commonly used to add flavour and aroma to dishes during the Ming Dynasty.
- During the late Qing Dynasty, tomato sauce, curry powder and other foreign cuisine condiments were introduced into the country.

4.4 The Emergence of Regional Cuisines

During the middle and late Qing Dynasty, diversified regional cuisines were firmly established. These regional cuisines were distinctive from each other due to factors such as availability of resources, climate, geography, history, lifestyle, and cooking methods.
The best known and most influential regional cuisines in China were Szechuan cuisine (Chuan), Shandong cuisine (Lu), Cantonese cuisine (Yue), and Jiangsu cuisine (Su).

4.5 The Establishment of a Complete Banquet Culture

Chinese banquets were fully developed during the Yuan, the Ming, and the Qing Dynasties where banquets were highly sought after. People developed a taste for extravagant and opulent banquet given the rapid socio-economic development and the cultural integration of all ethnic groups that were taking place at the time. Specialised banquets to suit a wide range of occasions were developed such as weddings, funerals, businesses, festival celebrations, extended family gathering etc. Capacity and capabilities of throwing a big banquet were observed. Complex and well-defined banquet etiquette were in place. Dishes served at banquets were finely cooked with beautiful taste and presentation.

Figure 3 Manchu and Han Imperial Feast

A good example was the Qiansou banquet (the banquet for a thousand elderly) initiated by Emperor Kangxi of the Qing Dynasty as a special celebration for his 60th birthday. More than a thousand men over the age of 65 attended and seated around 800 banquet tables.

In celebration of his 60th birthday, Emperor Kangxi held six banquets over three days with over 300 dishes, which consisted of at least 108 unique dishes from Manchu and Han Chinese cuisine and numerous snacks. The celebration was a symbol of the great integration between Manchu and Han Chinese ethnic groups. Officials from both ethnic groups attended the banquet together.

This 3-day banquet was called Manhan Quanxi (Manchu Han Imperial Feast). It was arguably one of the grandest banquets that was documented in Chinese cuisine.

Section 5 The Contemporary Period

Following the collapse of the Qing Dynasty, Chinese food culture was classified as entering the Contemporary Period. Chinese food culture in a modern context differs distinctively from the traditional food culture in the following five ways.

5.1 Advanced Kitchen Appliances

- Traditional energy sources such as wood logs and coal is replaced by gasoline, natural gas, solar, liquefied petroleum gas, and electricity for their efficiency.
- Modern kitchen appliances such as kitchen rangers, microwave ovens, fridges etc. are available to every household. Cooking, nowadays, has become much more time efficient, hygienic, and convenient for people.

5.2 Industrialisation of Food Production

Industrialization of food production is reflected in the following two areas:
- It is a fact that many commercial kitchens have replaced manual operations with machines in the food production processes.
- More traditional food such as Baozi (teamed buns with filling), Jiaozi (dumplings), noodles etc. are produced in the manufacturing product line which makes standardised food production possible.

5.3 Food Sources Continue to Grow

Globalisation has shortened the distance between China and the rest of the world. Many quality cooking ingredients from various countries are available in China. The technological innovations in agriculture have contributed to the enrichment of food sources.

5.4 Exchange of Culinary Resources Nationally and Globally

Thanks to the ease of modern transportation, increased population flow and fast information dissemination, the exchange of culinary resources including cooking techniques and skills and food products are taking place in a national and global scale frequently. Competitions of cooking skills are organised regularly as a means to enable the development of culinary arts.

The availability of the internet and global cookery books have increased the resources for chefs to be able to find new ideas and learn new skills. Additionally, with the international movement of chefs and international trade shows, more and more ideas are being shared across the globe.

5.5 The Trend of Chinese Food Development

The modern Chinese food development is based on a central principle of "people-oriented, taste- guided and skill-based". The vision is for Chinese food to meet people's demand for a balanced, time- efficient, and pleasant diet, combining the benefits of manufactured food with handmade cooking. To offer a nutritionally balanced, flavoursome and enjoyable dining experience.

Chapter 2 The Characteristics of Chinese Food Culture

Chinese food culture refers to the techniques, science, and arts associated with the cultivation, production and consumption of food. It also refers to the customs, traditions, ideology and philosophies that developed centred upon food. Chinese cuisine is food cooked in a Chinese style. It reflects the traditions of China as a multi-ethnic country and includes the essence of culinary arts of all ethnic groups.

Section 1 Characteristics of Chinese Food Culture

1.1 A Wide Range of Cooking Ingredients and Superb Culinary Skills

A Wide Range of Cooking Ingredients

China is a vast country with diverse landscapes. Officially, the country has seven geographical regions. However, traditionally people tend to consider the Yangtze River (the longest river in China) as a dividing line between the North Region and South Region. A summary of China's landforms and geographical features are shown below.

Table 3 Geographical Features of and Landforms

Geographical Regions	Main Geographical Features and Landforms
Northwest China	Sandy dunes, broad sunken basins and mountainous terrain with fertile lowlands and foothills
North China	Rolling prairies, plains, river valleys, hilly and mountainous terrain with lowlands and foothills
Northeast China	Forest, rolling plains and mountains with fertile lowlands and foothills
Southwest China	Snow-capped mountains, hilly terrain, and high plateaus with fertile basins

Chinese Food Culture

Geographical Regions	Main Geographical Features and Landforms
Central China	Plains, river valleys, lakes and hilly terrain with fertile lowlands and foothills and a vast lake system
East China	Hilly terrain, plains, and river valleys
South China	Hilly and mountainous terrain with deep river valleys and fertile lowlands and foothills on the southern coast

The diverse landforms provide the country with rich natural resources which in turn produce a wide range of cooking ingredients for Chinese cuisine to develop.

Cultivated food crops and livestocks including buffalos, pigs, goats, sheep, fowls etc., along with aquatic products, as well as wild plants and games are the main cooking ingredients.

Superb Culinary Skills

There are four aspects of the preparing and making of Chinese dishes that require skills.
1. selection of cooking ingredients
2. cutting techniques
3. seasoning techniques
4. control of the cooking temperature

The selection of ingredients is of significance to Chinese cuisine. Special attention is normally paid to the freshness, seasonality, the origin of produce, type, and different cut (as in meat) of the ingredients.

Chefs will apply desirable knife skills to cut or carve the ingredients into a variety of shapes such as shreds, slices, strips, cubes, segments, dices, wheat ears, leachy blossoms, and Suoyi etc. Regardless of the shape, ingredients need to be cut into pieces with relatively the same size and thickness to enable each piece to be cooked under the same heat at the same time.

There are dozens of cooking methods involved in making Chinese cuisine, for example, LIU(熘), BAO(爆), CHAO(炒), PENG(烹), DÙN(炖), MEN(焖), WEI(煨), WU(焐), JIAN(煎), YAN(腌), LU(卤), BASI(拔丝), GUASHUANG(挂霜), MIZHI(蜜汁).

There are also many different combinations of flavours including fresh and salty, salty and sweet, salty and spicy, mouth- fiery peppery (mouth-numbing) and spicy, sweet and sour, fragrant and spicy, Yuxiang and Guaiwei.

Depending on the nature of the cooking ingredients and the type of cooking method selected, chefs must be able to control the cooking temperature in a flexible way to enable dish to achieve the desired taste and texture.

Superior culinary skills are the foundation of the fame of Chinese cuisine.

1.2 A Wide Range of Cuisine Styles

Chinese cuisine varies based upon regional resources, varying climates, cooking methods, cultural differences, and religious food laws. Distinctive cuisine styles have evolved a long time ago.

In general, people from the South Region of China prefer to eat rice while people from the North Region tend to consume more wheat-based food.

In addition, the dominant taste of the regional cuisines from the North Region is salty, while that of the South Region, the East Region, and the West Region are considered to be sweet, spicy, and sour respectively.

The most well-known regional cuisines in China include Shandong cuisine in the lower reaches of the Yellow River, Sichuan and Hunan cuisine in the middle and upper reaches of the Yangtze River, Jiangsu and Zhejiang cuisine in the lower reaches of the Yangtze River, Guangdong and Fujian cuisine in the Pearl River basin, as well as integrated style of cuisines of Beijing and Shanghai.

In addition to the regional differences in cuisine among Han Chinese described above, each ethnic group in China has a set of unique traditions and practices associated with food, such as, Mongolian, Manchu, Hui (Chinese Islamic), and Tibetan etc.

Other vavirations of Chinese cuisine include the Imperial Court Cuisine, the Official Court Cuisine, the Temple Cuisine that is regulated by Buddhism and Taoism dietary laws, and the Folk Cuisine (working-class cuisine) in comparison with the Imperial and Official Court Cuisine.

Below is a nonexclusive list to classify Chinese cuisine by region, ethical group, and socio-economic class.

Table 4 Classification of Chinese Cuisine

By Region	By Ethnic Group
Shandong	Han
Sichuan & Hunan	Manchu
Jiangsu & Zhejiang	Mongolian
Guangdong & Fujian	Hui (Chinese Islamic)
Beijing	Tibetan
Shanghai	
Tianjin	
By Socio-economic Class	
Imperial Court	
Official Court	
Folk	
Temple	

Each style of Chinese cuisine is unique in its own way. Altogether, they form up the food culture shared by Chinese people.

1.3 A Manifestation of Chinese Culture

Chinese people have a long tradition of enjoying food as well as pursuing a gratified state of mind associated with food. The meaning of food to Chinese is far beyond its literal meaning.

The metaphorical meaning of food is well recognised. For example, Lao - tze (? - 531 BC), the founder of Taoism school of philosophy believed "ruling a big country is like cooking a small fish" which implies that too much handling will spoil a country in the same way that a small fish was spoiled.

Confucius (551 BC – 479 BC) philosophy insists that the wellbeing of a society depends on individual ethical behaviors. The code of conduct for junzi (gentlemen) includes "there is no objection to his rice being of the finest quality, nor to his meat being finely minced". This implies a gentleman's constant pursuit for perfection.

In addition, the symbolic meaning of food is widespread. For example, it is believed that pears should not be shared with relatives and friends because the pronunciation of pear (梨, phonetically spells as Li) in Chinese is the same as that of "separate" (离, phonetically spells as Li). Sharing a pear, therefore, bears a bed meaning of being separated unwillingly. Similarly, fish in Chinese, Yu, is a homophone for "affluence" or "abundance". This phonetic similarity has led to fish becoming a symbol of abundance and prosperity in Chinese culture. One of the most followed Chinese New Year (Spring Festival) traditions is to serve fish dishes at the meal of New Year's Eve. The fish dishes normally will be kept until the New Year's Day to wish for a prosperous new year.

It is fair to say that Chinese social norms, philosophical thinking, and traditions are manifested through food.

1.4 Combined Effect of Food and Medicine for Yangsheng (Nourishing Life)

Food Therapy

The unique Chinese tradition and system of food therapy were developed based on the concept that food and medicine share the same root. Chinese food therapy is a modality of traditional Chinese medicine whereby natural foods are used instead of medications for healing.

Chinese people have studied and practiced the effects of food on human organism for thousands of years. It is common for traditional medicine practitioners to prescribe food therapy based on its medical value while for chefs to consider the therapeutic effects and taste of ingredients to make nutritious dishes with healing effects.

In ancient times, the imperial healthcare system consisted of both food and medical departments to provide advice and services on healing and nourishing people's lives. Many of the Chinese traditional medical books include chapters on food therapy such as the ***Yellow Emperor's Canon of Internal Medicine***, ***Compendium of Materia Medica***, ***Dietetic Materia Medica*** etc.

Yangsheng (Nourishing Life)

Yangsheng is a healthy lifestyle in which a variety of self-cultivation practices such as Qigong (exercises), meditative, medicinal, and dietary practices are maintained to balance the body, mind, and spirit. The ultimate aims of Yangsheng are personal health and longevity. Dietary practice is an important aspect of Yangsheng as people believe there is a direct correlation between the food that people consume and their health. Thus, food is of great significance to Chinese people.

Although Yangsheng was considered a lifestyle exclusive to the upper class in the past, the concept of Yangsheng has permeated every aspect of life in China and has been affecting people's daily diet regardless of their social class. Chinese food therapy is undoubtedly a significant part of Chinese food.

Section 2 The Guiding Ideology of Chinese Traditional Diet

The guiding ideology of Chinese traditional diet comprises the following three ideals. The harmony of nature and human; Yangsheng (Nourishing life); harmonious blending of the five flavours of sweet, sour, bitter, piquant and salty.

2.1 Contents and Embodiment

Harmony Between Nature and Humans

For Chinese people, living in harmony with nature is a necessity. This includes adapting to nature in general and to their specific habitats in order to survive and nourish their lives.

Informed by this concept, the principle of selecting cooking ingredients is to consider not only the common basic requirements of the human body but also the specific needs caused by nature and bespoke habitat. That is why Chinese people follow the tradition having varied foods according to their geographic environment and in line with the four seasons.

For example, Shanxi Province in the North China region produces a large amount of high-quality wheat and traditionally people mainly live on a wheat-based diet. Whilst, as an adaptation to a relatively cold climate and the higher level of saline-alkali in soils, people love using vinegar for cooking to balance out the human body's acid alkaline level.

Yangsheng and Food Therapy Informend Nutrition System

As mentioned in Section 1 of this chapter, dietary practice is a significant aspect of Yangsheng. The Yangsheng dietary practices include:

1. using food instead of medication for healing
2. balancing diet according to an individual's body constitution
3. consuming food in balance

According to Chinese food therapy theory, food possesses medical value and is categorised by four properties including flavour, temperature, organ system to act on, and energetic action. The watermelon's properties are exemplified in the table below:

Table 5 Watermelon's Properties and Medical Values

	Watermelon Properties and Medical Values
Flavour	Sweet
Temperature	Cold
Organ system to act on	Heart, stomach, and bladder
Energetic action	Quenches thirst and relieves irritability, promotes urination and relieves heat
Medical values	Decreases the risk of obesity and overall mortality, diabetes, and heart disease. Soothes mouth ulcers and pharyngitis

Virtually every common foodstuff has been described in Chinese food therapy according to their unique properties. A connection between the five flavours of sweet, sour, bitter, piquant (spicy) and salty and the nutritional needs of the five major organ systems, the heart, liver, spleen, lungs and kidneys was also established. Based on an individual's body constitution, food should be carefully chosen to maintain a well-functioning body, free of diseases and full of energy to enable personal health and longevity.

The notion of consuming food in balance is normally considered from the following three aspects:
- Firstly, the amount of food and beverage taken should be in compliance with the amount of food a human body needs. Neither eating too much nor eating too little is good for health.
- Secondly, eat a diet that includes the right type of food for personal health. That is, people should eat a nutritious diet and should not be a picky eater.
- Finally, a greater variety of foodstuffs should be taken. Eating the same type of food too much creates imbalance in the human body. Also, the diet should be in accordance with the seasonal temperature, otherwise it will be harmful to health. Watermelon, for example, is cold and it acts to relieve heat. Therefore, it is a great summer fruit. However, taking it during wintertime may have adverse effects on the human body to certain people.

Harmonious Blending of the Five Flavours

Blending of the five flavours to create tasty dishes is a key Chinese culinary concept. It refers to the creation of new flavours through combining flavours of different ingredients as well cooking the ingredients in a variety of ways.

Chinese dishes are usually made up of the main ingredients, supplementary ingredients, and seasonings. The most common cooking method is stir-frying whereby flavours of all ingredients are fully integrated by using wok spatula to stir and by adding thickening agent to "glue" them together to enhance the integrated flavour.

The "blending" feature of Chinese dishes makes Chinese cuisine hard to describe, difficult to master yet wonderful to taste.

From a taste perspective, the process of "blending flavours" is used to transform simply flavours to a range of complex flavours, and to highlight the natural flavour of food or make savourless food tastier by seasoning and heating.

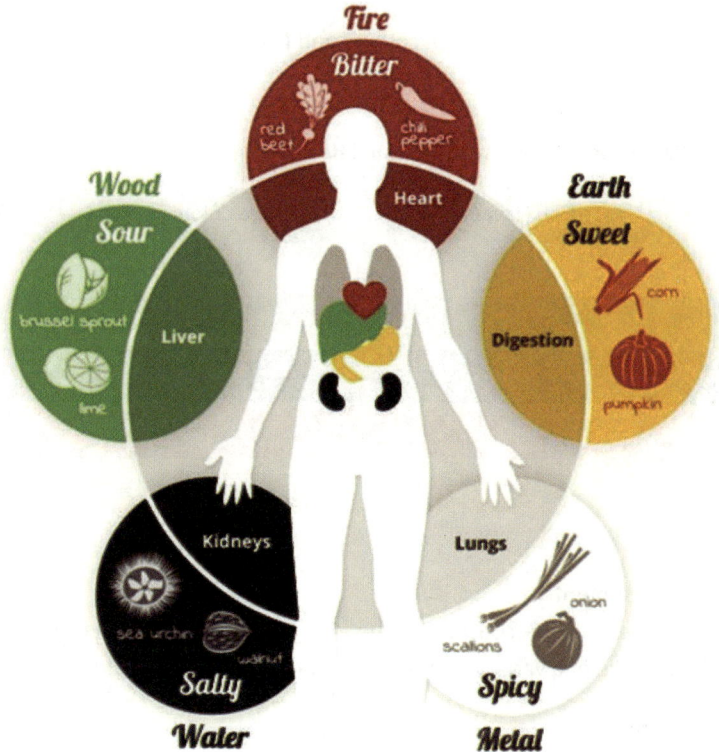

Figure 4 The connection between the five flavours and the five major organ systems
Image source: http://thespicedoc.com/content/glossary and designed by Patricia Callision

2.2 The Development of Food Science in Modern China

Traditional Chinese food science has developed over thousands of years. In modern times, further progress has been made and is mainly reflected in concepts concerning nutrition. Western thinking around a balanced diet is now accepted. Therefore, food science in modern China has developed to include both traditional Chinese food therapy and western nutrition.

Section 3 Traditional Chinese Diet Composition

Diet composition refers to food classifications and their relative proportions in the diet.

3.1 Traditional Chinese Diet Composition

There are four food classifications in the Chinese diet: grains, vegetables, fruits/nuts, and animal products. Their nutritional contributions were first recorded in the Yellow Emperor's Classics of Internal Medicine written in 2,600 BC as "The five grains provide nourishment. The five vegetables provide filling. The five domestic animals provide enrichment. The five fruits provide support." When balanced will, food will sustain life and maintain health.

Although these concepts were never stated to be dietary guidelines, Chinese people have been following them for over 2,000 years. This is especially the case for Han ethnic group.

The concept of "five" in terms of "five grains", "five domestic animals", "five fruits", and "five vegetables" is corresponding to the five-element theory. It refers to the five-representative food of each food group based on their food properties such as flavour, temperature, and the main human organ systems on which they have effect (as exaplained in Section 2 of this lesson). These terms were literally translated from Chinese to English for the convenience of this book.

The connection of five elements, five flavours, five colours and five main organ systems is presented in the table below.

Table 6 The Connection of Five Element Theory with Food and Health

Five Elements	Five Flavours	Five Colours	Five Main Organs
Metal	Spicy	White	Lungs
Wood	Sour	Green	Liver
Water	Salty	Black	Kidneys
Fire	Bitter	Red	Heart
Earth	Sweet	Yellow	Digestion

The Five Grains Provide Nourishment

Grains including cereals and beans were considered the vital source of nutrition to maintain life and health. It was advised to keep a daily intake of a variety of grains to obtain essential nutrition.

The Five fruits Provide Support

In ancient China, the common "five fruits" were peach, plum, apricot, chestnut and jujube. Nowadays, it referes to all kinds of fruit and nuts including dried fruit. Fruit was regarded as complementary food to grains, meat and vegetables to maintain good health.

The Five Domestic Animals Provide Enrichment
The five domestic animals include poultry, livestock, and their by-products such as dairy and eggs. Traditionally, Chinese people believe animal products are beneficial for human body to grow. However, they should be eaten in moderation i.e. in an appropriate amount and in ways that meet the needs of the human body to sustain good health. And, they should not replace the essential grains to become the principal food in the diet.

The Five Vegetables Provide Filling
The five vegetables include cultivated and wild vegetables. They are believed to provide complementry nutrients needed for a well-functioning body. Along with the essential food (grains) and the beneficial food (animals), a certain amount of vegetables can enhance people's immune system to prevent illness.

3.2 The Validation and Limitation of Chinese Traditional Diet Composition

The Validation of Chinese Traditional Diet Composition
A traditional Chinese diet is in line with the need for a healthy diet to keep fit. As modern dietetics pointed out , human body needs food to provide with seven vital nutrients such as carbohydrates, proteins, fat, mineral salt, vitamins, dietary fibres and water to sustain life and remain healthy.
Studies on the components of the traditional Chinese diet suggest that the essential food of grains and beans provides a large amount of carbohydrates and plant proteins meeting human bodies' basic needs for energy and proteins. In addition, "five animal products", "five vegetables", and "five fruits" provide complementary nutrients that grains cannot provide for, including animal proteins, fat, mineral salt, vitamins, dietary fibres and water.
In conclusion, the traditional Chinese diet meets the basic nutritional needs to sustain life and maintain health by offering exactly the seven essential nutrients.

The Limitation of Chinese Traditional Diet Composition
The limitation of traditional Chinese diet lies in its ambiguity and randomness. There are qualitative standards i.e. the varieties and quality of food were described but no specifications were given on the quantity.
Dieticians and medical scientists in Chinese history failed to put forward the specific quantitative criteria to decide on the quantity and proportions of food in diet, resulting in random amounts and portions of food when matching the food choices. This practice, to a certain extent, might have hindered the nutritional effectiveness of Chinese traditional diet.

Chapter 3 The Classification of Chinese Cuisine

Section 1 The Classification of Chinese Cuisine from a Historical Perspective

As we discussed in the previous chapters, Chinese cuisine is an important manifestation of Chinese culture. Chinese cuisine has developed in line with Chinese culture and has become hugely diversified. Depending on the regional preferences for seasoning and cooking techniques and cultural differences in ethnic groups, religious food laws and social classes, Chinese cuisine can be classified by regions, by ethical groups, and by socio-economic groups from a historical perspective. In addition, the category of Marketplace cuisine highlights the importance of catering industry to the development and integration of Chinese cuisine.
In this chapter, we are going to describe in detail twelve types of cuisine as listed below.

Part One

Table 7 Twelve Chinese Cuisine Styles

By Socio-economic Class
Folk Cuisine
Imperial Court Cuisine
Scholar-officials Family Cuisine
Temple Cuisine
By Region
Shandong Cuisine
Sichuan cuisine
Jiangsu cuisine
Guangdong cuisine
Beijing Cuisine
Tianjin
By Ethnic Group
Ethnic Cuisines

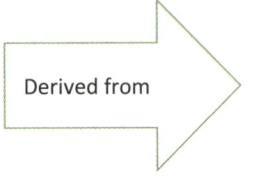
Derived from

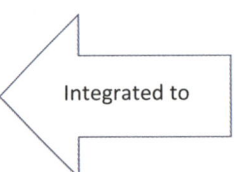
Integrated to

Marketplace Cuisine

1.1 Folk Cuisine

Folk Cuisine refers to the dishes prepared and eaten by common families in cities and rural areas from generation to generation. It includes everyday dishes and family feast dishes. Home cooked dishes are served three times a day, mainly including vegetable dishes and a few meat dishes. Everyday dishes are inexpensive and good for health. Family feast dishes are prepared for festival celebration and family gathering. Sumptuous family feast mainly comprises meat dishes with a few vegetable dish choices, and it is known for its generosity and approachable atmosphere.

1.1.1 History of Folk Cuisine

Folk Cuisine originated and developed among common families. It has strongly influenced cuisine from all over China. It is regarded as the base and source of every style of Chinese cuisine. Therefore, to a certain extent, folk cuisine is the root of Chinese cuisine.

1.1.2 Characteristics of Folk Cuisine

1. Simple ingredients and easy cooking techniques

    For ingredients of everyday meals, people usually use inexpensive local produce rather than luxurious ingredients. In terms of cooking techniques, there are no fixed rules. Simple and straightforward

cooking methods are chosen based on available ingredients and kitchen utensils.
2. Palatable taste without extravagant configuration
Seasoning of home-made dishes is designed to meet the taste preference of common families. Regional cuisine normally exemplifies the taste preference of people living in a certain region. Presentation was not the focus of home-made dishes. It is a palatable taste that people enjoy the most. Some popular dishes made by indigenous methods have a unique taste, though they may look ordinary.

1.1.3 Representative Dishes

There are a substantial number of dishes belonging to the Folk Cuisine category, of which a list of samples representative dishes are as follows.
- Pickles and twice-cooked pork often with chili seasoning in Sichuan
- Stir-fried freshwater snails and fried glutinous rice ball with sesame in Guangdong
- Mashed soya beans stir-fired with vegetables in Shandong
- Drunken shrimps and wine preserved crabs in Jiangsu
- Plain boiled pork with pig blood curd, braised pork with vermicelli and steamed sticky millet buns with red bean paste filling in Jilin
- Quick-boiling fish soup in Hebei

Family feasts in various places also boast distinct features, such as farm feast in Sichuan and soup feast in Luoyang.

Sichuan Province

PicklesTwice-cooked pork with chilli

Guangdong Province

Stir-fried freshwater snailsFried glutinous rice balls with sesame

Part One

Shandong Province

Mashed soya beans stir-fired with vegetables

Jiangsu Province

Drunken shrimps
Wine-preserved crabs

Jilin Province

Plain boiled pork with black pudding
Braised pork with vermicelli

Steamed sticky millet buns with red bean paste filling

Hebei Province

Quick-boiling fish soup

1.2 Imperial Court Cuisine

The Imperial Court Cuisine was created exclusively for royal families in ancient and imperial China. It was said to have represented the finest culinary skills and techniques of a dynasty.

1.2.1 History of Imperial Court Cuisine

Imperial Court Cuisine originated around the Zhou Dynasty (1046 BC - 256 BC). Imperial feasts served at the court of the Tang Dynasty were prepared with a very high level of culinary skills. Imperial banquets of the Southern Song Dynasty (1127 AD-1279 AD) included many extravagant meals. The quality and quantity of the Imperial Court Cuisine and the banquet etiquette had achieved sophistication during the Qing Dynasty.

1.2.2 Characteristics of Imperial Court Cuisine

1. Selection of the finest cooking ingredients
   As those who enjoyed royal cuisine owned a high social status, the choice and use of ingredients in royal cuisine maintained an extremely high standard. Any careless mistakes could be fatal for the chefs involved.
2. Exhibition of the highest level of culinary skills in a dynasty

Part One

The appointment of imperial chefs normally took a few rounds of screening. They were highly specialised and often possessed unique culinary skills. The finely equipped imperial kitchens provided the best conditions for cooking. A complex and complete management system for imperial cuisine was developed to regulate closely the style and components of royal diet, the selection and processing of cooking ingredients, the visual presentation, texture, and nutrition of dishes, tableware and even the name of the dishes.

3. Innovation in culinary skills and techniques

Two factors contributed to the constant improvement and innovation of Imperial Court Cuisine. Firstly, Emperors were fussy about food. It was the chefs' responsibility to meet emperors' demands by creating and improving imperial dishes. Secondly, imperial officials took the chance to please emperors by introducing new dishes.

1.2.3 Representative Dishes

The Imperial Court Cuisine took in many styles and accumulated a wide selection of dishes over the past thousands of years. Among the survived a few, the imperial cuisine of the Qing Dynastic is the most represented. Fang Shan and Yu Shan are the most well-known restaurants specialising in the imperial cuisine of the Qing Dynasty. Both located in Beijing, they produce acclaimed dishes such as:

Fermented soya bean paste stir-fried with cucumber, green peas, carrot, and hazelnut.

Luohan prawns

Chinese perch with eggs

Sweet pea cakes

Steamed cornmeal buns

Kidney bean rolls

1.3 Scholar-officials Family Cuisine

The Scholar-officials Family Cuisine originated from the dishes created and consumed by the Scholar-officials class. It had an emphasis on the health and hygiene aspects of food and was usually considered rather innovative in cooking techniques. Many dishes of this category enjoy a high popularity even to this day. It is fair to say that the Scholar-officials family cuisine reflected the extravagant lifestyle of scholars and high- ranking government officials at the time.

Part One

Kong Family Residence(Confucius Family Residence, 孔府)

1.3.1 History of the Scholar-officials Family Cuisine

The Scholar-officials Family Cuisine initially developed during the Spring and Autumn Period and started to take shape during the Han Dynasty and the Tang Dynasty. There was a further development from the Song Dynasty onwards. In addition to the long-established Confucius Family Dishes, there were famous family dishes in every dynasty.

Though the cuisine was designed to meet the demands of Feudal Scholar-officals class, it has made contributions to the development of Chinese cuisine by preserving the essence of Chinese food culture as well as improving culinary theories and cooking practices.

Tanjia Cai (Tan Family Cuisine，谭家菜)

Zhili Official Family Cuisine (直隶官府菜)

Kong Family Cuisine (Confucius Family Cuisine)：Two Phoenixes hatching from one Egg(孔府菜：一卵孵双凤)

1.3.2 Characteristics of Scholar-officials Family Cuisine

1. A wide range of cooking ingredients

    Given the privileged socio-economic status of scholars and high-ranking government officials, it was possible and affordable to obtain a wide range of ingredients with diverse qualities.

2. Ingenious culinary skills

    A government official's personal chef didn't have to have exclusive culinary skills to be selected compared with the recruitment methods used for imperial chefs. However, more often than not, they were more likely to develop ingenious skills, such as the skills demonstrated in a well-known dish of Tanjia Cai (Tan Family Cuisine), "bird nest in clear soup" (清汤燕窝).

Bird nest in clear soup

3. Elegant and Engaging Names of the Dishes

    Naming of their family dishes was of significance to government official families. Dishes were often given an elegant and engaging name to reflect the family status and tradition. Taking Kong Family Cuisine (Confucius Family Cuisine) as an example, "Shrimps Wearing Jade Belts" indicated the high status of the Duke of Yansheng and Kong Family; "Four Treasures of Study" indicated the scholarly tradition of the family.

Part One

Shrimps Wearing Jade Belts (翡翠虾环)   Four Treasures of the Study (文房四宝)

1.3.3 Representative Dishes

With the longest history and highest standards of its kind, Kong Fu Cai (Confucius Family Dishes) is the most distinctive example of Scholar-gentry family cuisine. It is rooted in the Lu Cai (Shandong Cuisine) and contains "everyday dishes" and "banquet dishes". Dishes such as "Yipin Pot", "Squab in the Duck", and "Poem, Rites and Gingko" are well known.

Kong Family Yipin Pot (孔府一品锅)   Squab in the Duck (带子上朝)

Poem, Rites and Gingko (诗礼银杏)

1.4 Temple Cuisine

Temple Cuisine, also known as vegetarian cuisine or religious cuisine, refers to dishes served in Taoism and Buddhism temples.

1.4.1 The History of Temple Cuisine

Following the increase of national productivity, vegeterian cuisine appeared first in the early Qin period. It grem rapidly during the Southern and Northern Dynasties which resulted in the establishment of Temple Cuisine. During the Tang and Song Dynasties, Temple Cuisine experienced accelerated development. At the time of the Qing Dynasty, Chinese vegetarian cuisine started to diverse to include three distinctive types, namely royal vegetarian cuisine, temple vegetarian cuisine and marketplace vegetarian cuisine.

1.4.2 Characteristics of Temple cuisine

1. Simple source of ingredients

   Most temples were built by mountains. In their daily life, in addition to doing Buddhist or Taoist services, monks, nuns, Taoist priests also spent a lot of time farming. A large amount of food could easily be obtained, and they made full use of these mountain resources.

2. Expert at cooking vegetables and beans

   Due to dietetic beliefs and religious disciplines, food used for temple cuisine mainly included raw plants such as fruit, bamboo shoots, fungi and bean products. After a lengthy period of practice, a good level of skills for cooking vegetables and beans were obtained.

3. Mimicking meat dishes with vegetarian ingredients

   Given the limitations in the vegetarian diet, improving cooking techniques was a means by which Temple Cuisine enriches its variety of dishes. Special efforts were made in the design of dishes, including mimicking meat dishes with vegetarian ingredients. For example, melon or chard along with eggs, salt, rice flour, bean flour and wheat flour were used to make "vegetarian pork", and soya bean curd sheet rolls were made into "shredded meat".

Gongbao Vegetarian Pork Belly (宫保素三层肉)

1.4.3 Representative Dishes

There is a wide variety of temple dishes. The most representative one is the "Luohan Zhai "(Buddha's Delight,罗汉斋). Some temple dishes use vegetarian ingredients to create the shape and texture of meat dishes. They often were named after meat dishes as well such as "vegetarian chicken"(素鸡), "boiled dried scallops" (白烧干贝)and "steamed soft-shelled turtle in crystal sugar soup" (冰糖甲鱼). They are so real that one could not tell the spurious from genuine meat dishes.

Luohan Zhai (Buddha's Delight,罗汉斋)

Part One

Steamed soft-shelled turtle in crystal sugar soup (冰糖甲鱼)   Boiled dried scallops (白烧干贝)

1.5 Ethnic Cuisine

The ethnic cuisine refers to cuisine made by 55 Chinese ethnic minorities apart from the Han nationality. These ethnic minority groups represent a decent proportion of the whole population of the country.

The History of Ethnic Cuisine

Ethnic cuisine developed in line with ethnic history. There have been more 55 minority nationalities living in the country since China became a republic in 1911, among which are Manchu, Mongolian, Korean and Hui (Muslims). Many minority ethnic dishes enjoy high popularity in China, such as "pork belly hot pot with pickled Chinese leaves" (酸菜白肉火锅), "grilled whole lamb" (烤全羊) and "hand-grabbed rice"(手抓饭).

Grilled whole lamb (烤全羊)   Hand-grabbed rice (手抓饭)

Pork belly hot pot with Pickled Chinese Leaves
(东北酸菜白肉锅)

1.6 Marketplace Cuisine

The Marketplace Cuisine, also known as restaurant dishes, refers to dishes made and sold by restaurants at marketplaces, acting as the main source of Chinese dishes.

1.6.1 The History of Marketplace Cuisine

According to historical data, the catering industry was initially formed during the Shang Dynasty. In the Tang Dynasty, there was a boom in the development of agricultural production, commerce and transportation. Throughout the Ming and Qing Dynasties, the characteristics of marketplace cuisine had become more distinctive, and a range of regional cuisines eventually became known.

1.6.2 Characteristics of Marketplace Cuisine

1. A collection of various cuisines with diversified cooking techniques
   Dishes of the marketplace cuisine derived from various regional cuisines and ethnic cuisines which makes Marketplace Cuisine rich in variety and cooking techniques.
2. Highly flexible and adaptable to changes
   Customers of the Marketplace Cuisine are from different regions and with diversified social backgrounds. Faced with fierce competition, the development of dishes sold by restaurants adapted to the changes of taste to meet different demands. The popularity of western food, the integration of local cuisines and the trend of consuming healthy food are examples that demonstrate the Marketplace Cuisine is flexible and adaptable to the changing of times.
3. Numerous types of flavours and distinctive styles
   To survive the fierce competition within the catering industry, a distinctive style is essential. Every restaurant would try its best to innovate new varieties of dishes to enhance their speciality, as well as to meet the demand of markets.

Part One

Adjusted version of duck soup with pickled radish (Confucius Family Cuisine, 神仙鸭汤)

Coriander Cake (芫荽饼)

1.6.3 Representative Dishes

Usually, many famous dishes popular at markets will eventually be integrated into the category of regional cuisine and even become the representative dishes of the region. Some of the most well-known examples will be given in the next section.

Section 2 Schools of Chinese Cuisine

China boasts a vast territory. Due to differences in natural conditions, local resources, habits and customs, economic and cultural development, distinct regional cuisines have emerged. Those famous dishes have their own unique history, reflecting excellent cooking skills. The stories or tales behind them are lovely and touching.

In this section, we will look at 5 most celebrated regional culinary traditions, namely Sichuan Cuisine, Shandong Cuisine, Jiangsu Cuisine, Guangdong Cuisine, and Beijing Cuisine.

The Great Four Styles of Cuisine and Beijing Style Cuisine.

2.1 Sichuan Cuisine

Brief Introduction of Sichuan Cuisine

Sichuan is famous for its cuisine, known by the Chinese as "to taste in Sichuan". Sichuan Cuisine, also known as Chuan Cai, is one of the most distinctive regional cuisines in China. It originated from the ancient Ba State and Shu State of China. By the late Qing Dynasty, it had fully developed to include a set of well-formed and unique culinary skills and a special local flavour with strong features. It is regarded as one of the "Four Great Cuisines" of China along with Shangdon Cuisine, Jiangsu Cuisine and Guangdong Cuisine. Its influence has gone beyond the country's borders.

Components of Sichuan Cuisine

Sichuan Cuisine mainly includes Chengdu style and Chongqing style.

- Chengdu style emphasises the selection of ingredients, knife skills, seasoning and heat control. The central characteristic of Chengdu cuisine is its deep and strong flavour, as well as dishes that tend to be salty or spicy.
- Chongqing style chefs have a tendency of not sticking to established practices. The widespread tradition is to create new recipes which lead to the birth of many new styles of dishes. Usually, it is grass root chefs who are more likely to create new dishes, also known as grass root dishes.

Features of Sichuan Cuisine

Sichuan style cuisine is known for its variety of flavours. As the saying goes, "one dish with one flavor; with one hundred dishes, it comes with hundred flavors." The five primary tastes of Sichuan cuisine include fiery peppery (mouth-numbing), pungent, sweet, salty and sour.

Sichuan Cuisine pays attention to the variation of flavours. It differentiates between "deep and light" as well as "strong and mild". Flavours can be adjusted according to the preference of consumers or the change of seasons. However, some Sichuan dishes would never adjust in order to preserve their traditional flavour, such as Chongqing "beef tripe hot pot". It will be served as usual in summer, with no adjustment to its tastes and the way in which it is consumed. People refer to this tradition as "resisting heat with heat".

One of the most outstanding characteristics of Sichuan Cuisine is the skillful balance between tastes in seasoning. Strong and mild tastes are brought out in an orderly manner so that they are separated yet perfectly blended to make flavours rich but not greasy, light but not plain. As a result, Sichuan Cuisine is known for its intense flavours. On one hand, on the other hand, its milder dishes are impressive as well. The main types of flavour of Sichuan Cuisine include "Mala" (mouth-numbing and spicy), "Yuxiang" (fish fragrant), "Jiachang" (home style), "Guaiwei" (strange/exotic taste), "Suanla" (sour-spicy), and "Tangcu" (sweet-sour). Among them, "Yuxiang" (fish fragrant), "Jiachang" (home style), and "Guaiwei" (strange/exotic taste) were invented by Szechuan chefs.

Representative Dishes

Gongbao Diced Chicken (宫保鸡丁)

Gongbao Chicken, one of the most acclaimed traditional Sichuan dishes, is now widely served in China. The story suggests that this dish was named after the provincial Governor of Sichuan in the Qing Dynasty, Mr. Baozheng Ding (his courtesy name was Gongbao).

Governor Ding was a gourmet. During his office in Shandong, he hired ten famous chefs. When he moved to Sichuan, he brought with him several family chefs of his own. Fast stir-fired diced chicken was always served at Ding's banquets. It was highly appreciated by all his guests for its umami and aromatic flavour. However, none of them were able to replicate it at home. This dish was referred to by people who had eaten it as "Gongbao Diced Chicken".

Gongbao Chicken (宫保鸡丁)

Part One

Mapo Tofu ("Pockmarked Granny" Bean Curd, 麻婆豆腐)

Mapo Tofu is another famous Sichuan dish. It was said that in the Tong Zhi Period (1856 – 1875) of the Qing Dynasty, there was a small family-run restaurant near Wanfu bridge outside the north gate of Chengdu city of Sichuan. The restaurant owner was named Chen. She was very good at cooking. One of her specialties was to stir-fry tofu in with chili, Sichuan pepper, fermented soya bean paste, and often topped with minced meat to create a combination of fiery peppery (mouth-numbing), spicy, umami, and aromatic tastes. It was delicious and very popular.

This dish did not have a name back then. Chen was an elderly woman with pockmarks on her face. Her customers started to call this dish "mapo tofu" as mapo in Chinese literally means pockmarked granny.

Other typical dishes of Sichuan Cuisine include "Fish-fragrant shredded pork"(鱼香肉丝), "Twice-cooked sliced pork belly"(回锅肉), "Water boiled beef"(水煮牛肉).

Mapo tofu (麻婆豆腐)

Water boiled beef (水煮牛肉)

Fish-fragrant shredded pork (鱼香肉丝)

Twice-cooked sliced pork belly (回锅肉)

In term of snacks, there are "Zhong style dumplings" (钟水饺), "Dandan noodles" (担担面), "Long style chao shou" (龙抄手). Chaoshou is an unique way Sichuan people refer to wontons, which literally means making wontons in the shape of folded hands. "Lai style tangyuan" (glutinous rice balls, 赖汤圆), "North Sichuan style liangfen"(sweet peas jelly, North Sichuan style 川北凉粉) etc.

Zhong style dumplings (钟水饺)

Dandan noodles (担担面)

Long style chao shou (龙抄手)

Lai style tangyuan (glutinous rice balls, 赖汤圆)

North Sichuan style liangfen (Sweet peas jelly, North Sichuan style, 川北凉粉)

2.2 Shandong Cuisine

Brief Introduction of Shandong Cuisine

Shandong Cuisine is also known as Lu Cai. Given its special contributions to the development of Chinese food culture and Chinese culinary techniques, Shandong Cuisine is considered as a significant element of Chinese cuisine. The initial form of Shandong Cuisine appeared as early as the Spring and Autumn period. Since then, Lu Cai enthusiasts and professional chefs have been passing down the traditional techniques whilst continuously innovating and enriching the Shandon style by learning from the best practices of other cuisines in China. It is believed that Shandon Cuisine has reached a period of great prosperity.

Components of Shandong Cuisine

Shandong Cuisine includes three culinary styles, namely Ji'nan cuisine, Jiaodong cuisine, and Ji'ning cuisine.
- Jinan style tends to use a wide range of fine ingredients to create a featured flavour which can be described as light and aromatic, crispy and tender, thick and pure. Soup-making is a specialty of Ji'nan tradition. It is believed that cooking dishes in specially prepared broth can create a variety of flavours. The representative dishes include "Two-coloured squid rolls" (双色鱿鱼卷), "Basi apples" (Toffee apples, 拔丝苹果).

Part One

Two-coloured squid rolls (双色鱿鱼卷)   Basi apples (Toffee apples, 拔丝苹果)

- Seafoods are the main ingredients of Jiaodong style. To preserve the cut, colour, and taste of the ingredients, tastes of Jiaodong dishes tend to be light and mild, savoury and tender. Food presentations are important as well. Creating feasts on single-main-ingredient seafood is a famous tradition. For example, fish feast and sea-Crab feast are well known.

Fish feast

Sea-Crab feast

Chinese Food Culture

- Ji'ning style, also known as Qufu style, is represented by "Kong family cuisine" (Confucius family Cuisine, 孔府菜). Kong family cuisine emphasises all aspects of culinary tradition including the fine selection of ingredients, accurate control of heat, strictly managed cooking process, and the nutritional effects of the food to enable Yangsheng (nourishing life). Dishes of this style are mellow tasting that intensify the natural flavour of food. The presentation style of Kong family dishes is elegant and gorgeous to make it one of the high-end cuisine brands in China.

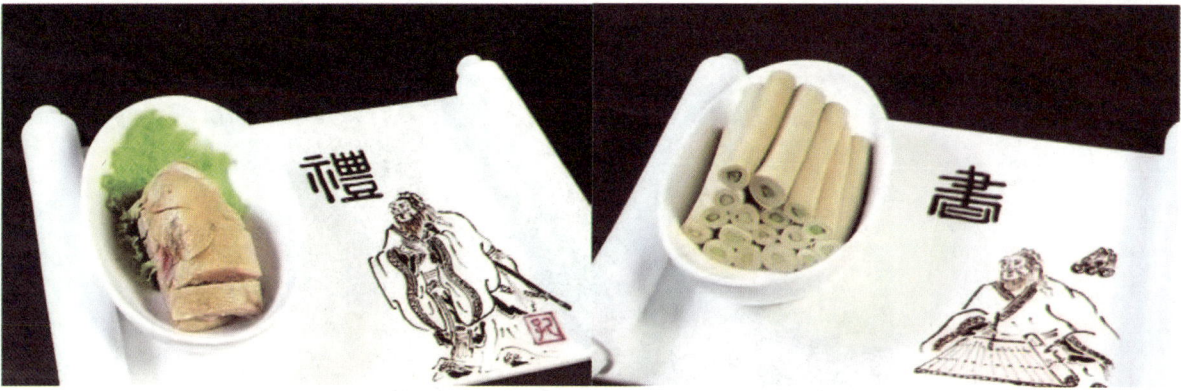

Features of Shandong Cuisine

Shandong Cuisine usually tastes salty and savoury. Chefs aim to bring out the natural flavours of food; therefore, the choice of seasoning is determined by the flavour of ingredients. Spring onions is a common herb used in Shandong dishes either as spice or a condiment. The aroma of spring onions is the signature of Shandong flavour. In addition, broth is commonly used in seasoning. The umami taste of seafood is emphasised in cooking. Shandong culinary tradition has developed a range of complex flavours such as "Five Spices", "Sour-spicy", "Sweet-sour" and "Sesame paste".

The tradition of consuming spring onions and garlic raw with of without dipping sauces is famous. Shangdong spring onions is famous in China for its juicy and crunchy texture, and very mild and fresh taste.

Representative Dishes

Dezhou Paji (Dezhou Braised Whole Chicken, 德州扒鸡)

Part One

Dezhou Paji has a long history and fame enjoyed nationally and internationally. It was invented in 1911 by Mr. Shigong Han and his colleagues. Han was the owner of Deshun Zhai (Deshun House, 德顺斋), a time-honoured restaurant specialised in making stewed chicken. He and his colleagues modified the traditional recipe and cooking techniques in the following two ways.

- Added in a few Chinese traditional medicine herbs which have the effect of nourishing spleen and stomach.
- Adapted Mr. Baoqing Hou's combined method of cooking chicken to including deep-frying, smoking, braising, and roasting.

After numerous experiments, they had finally succeeded the "Five-spice Chicken" in which the body shape of a chicken is kept but the meat can be easily separated from the bones.

In the new recipe, they had considered the taste of their region and also taken into account the preferences of people living in other regions of China, which are best described as "Sweet South, Salty North, Sour East".

Using this new combined method, chicken was deep-fried evenly, stewed to make it tender and aromatic, and able to be preserved for a long time. It did not take long for this dish to become a national favourite.

Guo Shāo Pork Shank

"Guo Shāo pork shank" is a famous hot dish in Shandong. It was derived from "Guo Shāo pork" which first appeared in the Yuan Dynasty. Pork shank (front legs) has gradually become the chosen ingredient of chefs. There are multiple cooking methods involved in making this dish, including boiling, steaming, and frying. In between the methods, knife skills are required to prepare the pork shank.

Guo Shāo pork shank

Other Famous Dishes and Snacks Include the Following:

Huobao Liaorou (sliced pork cooked in the wok over a big hot flame, 火爆燎肉)

Clam Soup (氽西施舌)

Steaming Abalone with Shells (原壳鲍鱼)

Fried Yuanyang Gazha (炸鸳鸯嘎渣)

Guota Tofu (锅塌豆腐)

Pan-Fried Stuffed Buns (水煎包)　　　　　　Zhoucun Sesame Crispy Bread (周村酥烧饼)

2.3 Jiangsu Cuisine

Brief Introduction of Jiangsu Cuisine

Jiangsu Cuisine has a long history. Throughout the Sui, the Tang, the Southern and Northern Song Dynasties, prosperous markets in Jinling city and Yangzhou city boosted the development of Jiangsu Cuisine. In the Qing Dynasty, cooking techniques of Jiangsu tradition had become more meticulous with some unique features and a wide variety of dishes. Its influence had spread throughout the whole country.

Jiangsu Province, with its fertile lands and a vast lake system is known as the "home of fish and rice". There are abundant resources all year-round, including a wide range of fruits and vegetables. All of which provide plentiful cooking ingredients to enable the development of Jiangsu Cuisine.

Components of Jiangsu Cuisine

Jiangsu Cuisine has four sub-region cuisine styles including Huaiyang style, Jinling style, Wuxi style and Xuhai style.

1. Huaiyang style cuisine is renowned for its light and mild flavour which suits the taste of people originated in both southern and northern China. It is the main component of Jiangsu Cuisine.

2. Jinling style dishes are well-known for their pure, mild, smooth, and mellowing tastes, of which Salty Duck is the most famous one. It is also representative dish of Nanjing duck dishes.

Salted Duck (咸水鸭)

3. Wuxi style cuisine is good at bringing out the natural flavour of ingredients. Its representative dishes include "Biluo shrimps" (碧螺虾仁), "Lake Taihu sliver fish" (太湖银鱼), "Jiaohua ji" (Beggar's chicken, 叫花鸡), "Squirrel-shaped deep-fried mandarin fish" (松鼠鳜鱼) etc.

Biluo shrimps (Shrimps with Biluo tea leave, 碧螺虾仁)

Lake Taihu sliver fish (太湖银鱼)

Squirrel-shape deep-fried mandarin fish (松鼠鳜鱼)

4. Marine products are the main ingredients of Xuhai cuisine. Dishes are cooked in a simple and practical style with salty and savoury taste. The typical dishes include "Farewell, my concubine" (soft-shell turtle & chicken soup, 霸王别姬) and "Red cooked goby fish" (红烧沙光鱼).

Farewell, my Concubine (soft-shell turtle & chicken soup, 霸王别姬)

Red cooked goby fish (红烧沙光鱼)

Features of Jiangsu Cuisine

Light and mild yet savoury is the key flavour of Jiangsu Cuisine. Food is prepared in such a way that the freshness and umami of the ingredients are highlighted. In addition, Jiangsu Cuisine tends to emphasise on appropriate blend of vegetables and meat ingredients to bring out the pure taste.

Chinese Food Culture

Representative Dishes

Salted Pig Trotters in Jelly (水晶肴蹄)

Salted pig trotters in Jelly, also known as Zhenjiang pig trotters, is one of the most famous dishes in Zhenjiang. The story behind this dish suggested that more than 300 years ago, there was a family-run restaurant in Zhenjiang city of Jiangsu Province. One day, the husband bought four pig trotters. He meant to prepare them in salt overnight to prevent them from going bad. However, he didn't realise that nitrate was mistaken for salt in the process. The next day, when he was about to cook the trotters, he noticed the texture of the meat had become firmer and even the colour of the meat looked fresher and cleaner. He soon found out his mistake. To remove the taste of nitrate, he soaked the pig trotters in water several times, boiled them, and rinsed them with fresh water again. Then he braised the trotters in a pot with spring onions, gingers, Sichuan pepper, cinnamon, fennel seeds for more than an hour. His initial plan was to detoxicate the trotters with heat. Unexpectedly, the braised trotters produced an extremely delicious aroma. Everyone who tasted the trotters was convinced that this dish was a real delicacy. Since then, this dish has become a chef's special of this restaurant. The dish was named "Salted pig trotters in jelly". It has become a representative dish of Zhenjiang ever since.

Jiaohua Ji (Beggar's Chicken, 叫花鸡)

Part One

Once up a time, there was a beggar who was hungry and cold on Chinese New Year's day. He stole a hen and killed and gutted it using bare hands. He seasoned the cavity with a pinch of salt and left it to marinate.

When he realised there was no pot or other utensils to be found, he decided to pack the yellow clay mud around the bird. He then set the bird in a hole that he dug on the ground where he had lit a fire. He buried the bird to cook it with the fire on top.

He fell asleep while he was waiting. After a long nap, he dug the chicken up and cracked open the clay. The chicken feathers were removed with the clay. He found the meat to be tender and aromatic. The unique aroma of the bird attracted a nearby restaurant owner who approached the beggar for the recipe. At the end of the story, the restaurant owner bought the beggar's technique. He added more spices to the recipe and the dish became very popular. The dish was named after the bagger "Jiaohua Ji". It literally means "Beggar's chicken" in Chinese.

Other famous dishes and snacks of jiangsu style include the following:
Dish

Yangzhou Lion Head (Yangzhou style meatballs, 扬州狮子头)

Liangxi Crispy Oriental Weatherfish (梁溪脆鳝)　　　　Wensi Tofu (文思豆腐)

Snacks:

Four-happiness Glutinous Rice Balls(四喜汤团)

Fengzhen Noodle (枫镇大面)   Guogai Noodle (锅盖面)

Huangqiao Sesame Shāobing (黄桥烧饼)   Zhuangyuan Dumplings (状元饺)

2.4 Guangdong Cuisine

Brief Introduction of Guangdong Cuisine

Guangdong Cuisine originated from the region around Guangdong, also known as Yue cai (Cantonese cuisine). It is one of the "Four Great Cuisines" in China. Located on the southern coast of China, the rich varieties of fruit and animal resources from the mountains along with marine and freshwater products contribute to the rich varieties of local food. Guangzhou, the capital city of Guangdong is a port city with a long and established history of international trading. Many Chinese who moved overseas have introduced the culinary traditions of European countries, American countries, and Southeast Asian countries back to the region, allowing Guangdong Cuisine to absorb the features of other cuisines.

Guangdong has a tropical and subtropical monsoon climate with long summers and plentiful rainfall. Surrounded by mountains and facing the sea, favourable geographical conditions have provided abundant resources necessary for the diversified food varieties in Guangdong since ancient times. In addition to the ordinary products of chicken, duck, fish and shrimp, Guangdong Cuisine also uses unusual ingredients such as snakes, cats, squirrels, snails and silkworm cocoon.

Components of Guangdong Cuisine

Guangdong Cuisine consists of Guangzhou style, Chaozhou style and Dongjiang style.

1. **Guangzhou style** Cuisine is the main component of Guangdong Cuisine and the fine representative of traditional Cantonese Cuisine. Its features include:
    - Diversified ingredients varieties, a wide range of dishes and condiments.
    - An emphasis on preserving the natural flavour of the food. Sometimes lighter than most other Chinese regional cuisines, Guangzhou dishes are prepared carefully and exquisitely to enable the fresh, crispy, tender, and smooth texture and mild flavour. Seasonings are varied which can be stronger and heavier in taste in wintertime.
    - A variety of cooking methods were developed but the most commonly used one is quick stirfry.

The representative dishes are listed below:

Chopped Plain Chicken (白切鸡)

Quick-boiled Prawns (白灼虾)   Steamed Seafood (清蒸海鲜)

2. **Chaozhou style** cuisine has absorbed some advantages of Western cuisine, making it a unique style among other Chinese cuisines. It is also the most developed style of Guangdong Cuisine specialising in seafood dishes. Famous dishes include Roast Goose and Chaozhou Stews.

Roast Goose(烧雁鹅)   Chaozhou Stews(潮州卤水)

3. **Dongjiang style** dishes are also known as Hakka cuisine. The ingredients mainly include poultry, livestock, and tofu products. Aquatic products are rare. Dongjiang dishes aim to be a smooth and soft, thich and aromatic flavour with a relatively salty and heavy taste. Usually, dishes require longer cooking time. A wide variety of one-pot dishes are available. The representative dishes include "Salt Baked Chicken" (盐焗鸡), "Crispy Duck" (香酥鸭), "Fried Slices Pork Belly Steamed with preserved vegetables" (梅菜扣肉).

Salt Baked Chicken (盐焗鸡)

Crispy Duck (香酥鸭)

Fried Slices Pork Belly Steamed with preserved vegetables (梅菜扣肉)

Features of Guangdong Cuisine

Guangdong Cuisine represents the typical southern China style flavours and tastes which are known for being fresh, crisp, tender, and smooth. Dishes consumed in Summer and Autumn tend to be mild and light while in winter and spring food becomes slightly richer and heavier. Guangdong Cuisine focuses on highlighting the original flavour of the ingredients.

People from Guangdong are known for having a discerning palate when it comes to the freshness of their meat dishes. To keep animal food ingredients alive until the moment of cooking is customary practice. There are specific requirements for heat control in the process as well. Cooking time is short to preserve the original flavours and nutrition of the food.

Varied sauces to go with dishes

Representative Dishes

Baiyun Trotter (白云猪手)

As the name suggests, there is a story behind this famous Guangdong dish. The story says there was a Buddhist temple on top of Baiyun Mountain. One day the abbot of the temple went on his alms round. Knowing it was against Buddhist food law, a young monk couldn't resist the temptation of eating meat. He managed to smuggle a pig trotter in as a treat. He boiled it secretly in a pottery vessel outside the main gate. Unfortunately, when the trotter was ready, the abbot made an early return. The young monk quickly ran to the nearby brook and threw the trotter in the water to cover-up and hide the trotter. The next day, a passing woodcutter discovered the trotter by chance. He brought it home and seasoned it with sugar, salt, vinegar etc. At this point, the trotter had become crunchy outside and crispy inside with a delightful sweet-sour taste. The method of cooking pig trotter became popular among locals soon. Since this technique originated in Baiyun Mountain, people call it Baiyun Trotter.

Part One

Other well-known dishes include "Fried-pork in sweet and sour sauce" (糖醋咕噜肉), "Dry-fried prawn jujubes" (干炸虾枣) and "Quick stir-fried perch meatballs"(香滑鲈鱼球). Examples of famous snacks are "Steamed bun stuffed with roast pork" (叉烧包), "Century egg Short Cake"(Pidan Short Cake, 皮蛋酥) , Cantonese style moon cake etc(广式月饼).

Gulao Pork (Fried Pork in Sweet and Sour Sauce, 糖醋咕噜肉)

Dry-fried Prawn jujubes (干炸虾枣)

Quick Stir-fried Perch (香滑鲈鱼球)

Cha Siu Bao (Steamed bun stuffed with roast pork, 叉烧包)

Chinese Food Culture

Century egg Short Cake (Pidan Short Cake, 皮蛋酥)

Cantonese Style Moon Cake (广式月饼)

2.5 Beijing Cuisine

Beijing Cuisine, also known as Jing Cuisine, refers to local dishes of the Greater Beijing area. Beijing Cuisine perhaps is the most complex one among all styles of Chinese cuisine. It is a combination of Shandong Cuisine, Chinese Muslim Cuisine, Imperial Court Cuisine and Scholar-officials Family Cuisine.

Beijing Cuisine is light, aromatic, umami, tender, and crisp with an emphasis on food presentation, balanced nutrition, meticulous cooking methods, and a wide range of flavours.

Part One

The most famous dish of Beijing Cuisine is the internationally renowned "Peking Roast Duck"(北京烤鸭). Regarded as a treasure of Chinese culinary art, the secret of Peking roast duck lies in the special type of duck that is selected. According to legend, the Peking duck is one of the best species in the carnivorous duck family. The late Premier Enlai Zhou was a real fan of "Peking Roast Duck". He had been to the acclaimed Quanjude Roast Duck Restaurant 29 times during his premiership for work or holding reception banquets to entertain his international guests.

Quanjude Roast Duck Restaurant （全聚德）

Chinese Food Culture

Among many dishes and snacks, the most popular ones include "Lamb Hotpot" (涮羊肉), "Grills" (烤肉), "Fried Fish Slices" (抓炒鱼片), "Stir-Fried Flour Knots" (炒疙瘩), "Shāobing with Pork Mince" (肉末烧饼) and "Steamed Cornmeal Buns" (小窝头).

Lamb Hotpot (涮羊肉)

Part One

Grills (烤肉)

Fried Fish Slices (抓炒鱼片)

Stir-Fried Flour Knots (炒疙瘩)

Shāobing with Pork Mince (肉末烧饼)

Steamed Cornmeal Bun (小窝头)

Chapter 4 Chinese Food Related Folk Customs and Etiquettes

As we learnt in the previously Chapters, Chinese food culture has developed alongside the nation's 5000 years of history and therefore is an important component of Chinese history. In a country with such a long history, wide territory and many ethnic groups, Chinese food customs and protocols vary with times, regions, and ethnic groups, contributing to a diversified food culture.

Section 1 Daily Food Customs

1.1 Key Features of Chinese Daily Food Customs

The importance of food to Chinese people has been captured in a 2000-year-old famous saying: "民以食为天" (Ming Yi Shi Wei Tian, meaning "Food is the first necessity of life"). Chinese people have long established the nutritional effect of food to maintain health. As a vast country, China boasts a wide range of food related customs. Each ethnic group has its distinctive food habits and traditions due to differences in living environment, historical development, and religious belief. It is those habits and traditions that formulate the complete system of Chinese food customs.

For example, in Southern China people consume more rice-based staple food due to a large quantity of rice cultivation. People in Northern China eat more flour-based staple food as the major crop grown in the region is wheat. Similarly, ethnic groups who rely on crop cultivation to make a living tend to consume grains and vegetables on a daily bases. For those whose economy is based on livestock production, meat and dairy products are their main food.

1.2 Daily Food Customs of Han Chinese

As we mentioned before, China is the home of 56 ethnic groups. The largest group is Han, making up 92% of the total population.

Han people reply on a largescale of crop farming and smaller-sized livestock production to provide stable food resources. Their diet is primarily crops supplemented with meat products and vegetables. Han people have developed a habit of eating three meals a day with lunch and supper as dinner.

Rice and flour products dominate Han people's staple food list. Corn, sorghum, grains, and potatoes are also commonly used. There are hundreds of ways to cook staple foods, for example, for rice there are steamed rice, rice cake, two kinds of rice dumplings (tangyuan and zongzi) and for wheat there are steamed bun (mantou), steamed stuffed bun (baozi), noodles, dumplings (jiaozi), all of which are popular daily food for Han people.

Part One

Han people have a long history of consuming alcoholic beverages and tea. China is not only the cradle of tea, but also one of the earliest countries to master alcohol-making techniques. There is evidence to suggest that the Chinese may have independently developed the process of distillation during the Eastern Han dynasty (25 AD - 220 AD).

Alcoholic beverages, including Baijiu (white wine) and Huangjiu (yellow wine), are an essential element of Han people's food culture. Baijiu literally means clear liquor is made from water, cereal grains such as rice, sorghum, millet, or wheat, and jiuqu (starter culture). It is a strong distilled spirit that contains a higher percentage of alcohol. Although the strength differs according to the brand and variety, Baijiu ranges between 8% and 56% alcohol by volume. Unlike Baijiu, Huangjiu is not distilled and contains less than 20% alcohol.

Famous Baijiu brands

Chinese Food Culture

Huangjiu

The symbolic serving and drinking of alcohol on various occasions and in different places can convey many meanings. For example, on the wedding day, the bride offers a dutiful toast to her parents before leaving home to thank them for their love and care. No wedding ceremony is complete unless the happy couple drink wedlock wine with their arms linked. It's a symbol to show that they become one and their future is intertwined.

Wedlock wine at traditional Chinese Wedding Ceremony

The Chinese traditionally serve Baijiu and Huangjiu either warm or at room temperature in a small ceramic bottle with a few small cups. Today glassware is a popular alternative.

Traditional Chinese Wine Set

59

Part One

In addition to alcoholic drinks, Han people have a strong tea culture. The tea-drinking tradition is integrated with health maintenance, cultural activities, and socialising, making tea the most widespread beverage, and tea-drinking the most common cultural activities in this country.

The core of Chinese tea culture is harmony. Drinking tea is regarded as a process of achieving harmony between tea leaves, water, temperature, tea wares, setting, and the mind of tea drinkers. It provides a situation, a scene and a formality whereby people cultivate and practice mindfulness, to find peace of heart and mind.

The health benefit of tea is another key aspect of Chinese tea culture. There is a wide range of tea products for people to choose from to suit their physical and mental needs.

6 major tea types

Chinese Food Culture

Section 2 Food Related Folk Customs in Festivals

Since ancient times, food has been a significant part of festivals and celebrations regardless of people's cultural background. In China, festival food customs were developed to strengthen family ties and bring balance to people's lives through which people express their aspirations, expectations and other psychological or cultural needs, as well as their aesthetic consciousness.

2.1 Chun (Spring) Festival

The Spring Festival (also known as Chinese New Year) is the most important traditional festival in China to celebrate the start of the lunar new year. It is an occasion for a family reunion. New Year's Eve (Chuxi) is also called "Reunion Eve".

The month prior to Chinese New Year is called "La Month", similar to Advent in Christianity. Xiao Nian Festival, literally means "Minor New Year Day", is the beginning of the Spring Festival season kicks off a series of preparations and celebrations of the Spring Festival hence it is considered the Prologue of the Spring Festival.

The whole begins from this day for and will last till the Lantern Festival on the fifteenth day of the first lunar month.

Many traditional folk customs are still in practice in the days leading up to New Year's Day.

There are indeed plenty of complicated dining protocols to follow. The joy of celebration is always associated with food.

Eating dumplings is an essential tradition. Even the poorest families make great efforts to prepare a feast of dumplings on New Year's Eve. Dumplings are eaten at the junction (交) of the old and new, specifically during the time from 11 p.m. to 1 a.m. (子时), Chinese put the two characters 交 (jiao) and 子 (zi) together to name that food as 饺子 (jiaozi). New Year rice cake (niangao) is also a must-eat food because it signifies making big-progress and fortune year after year.

Dumplings

Part One

Chicken, fish, sweet potato noodles are also the must-haves on the annual reunion dinner table because chicken (鸡ji) symbolises good luck (吉ji), fish (鱼yu) means abundance (余yu), and the long, sweet potato noodle represents longevity. In addition, apples and mandarin are on the list, signifying peace and auspiciousness.

Typical food served in the annual reunion dinner

Origin of "Sending the Kitchen God Off"

Spring Festival celebrations start with the ritual of sending the Kitchen God back to Heaven on the 24th day of the La Month. It is said that the Kitchen God is dispatched by the Jade Emperor from above to overlook the daily dynamics of the family.

The Kitchen God is accompanied by two guards, one holding a good jar and the other an evil jar. They are all ready to record the good and evil behaviour of the family and keep it in their jars accordingly. At the end of the year, the Kitchen God will return to Heaven to report to the Jade Emperor on what the family has done during the past year. Based on their behaviour, the family will be given blessing or adversity. Therefore, every household practices the tradition of offering the Kitchen God some sweet treats before sending him off. That story shows the self-discipline of the ancient Chinese people and their ethical pursuit, valuing the good and degrading the evil.

Melon Candy

2.2 Qingming (Tomb-sweeping Day) Festival

The Qingming festival is to worship our ancestors. It is known in English-speaking countries as Tomb-sweeping Day. It falls on the 1st day of the 5th solar term of the 24 solar terms according to the traditional Chinese lunisolar calendar. 24 solar terms were developed by ancient Chinese based on the changes of the sun's position in the zodiac throughout the year to guide farming practices. They indicate seasonal changes in the weather. People also follow them to prepare in advance to cope with weather changes and live a healthy lifestyle.

Twenty-Four Chinese Solar Terms		
24 Solar Terms	**Solar Date, on or around**	**Indications**
Spring		
Beginning of Spring	04 February	The spring season begins
Rain Water	19 February	The amount of rainfall increases
Insects Awaken	05 March	Insects are awakened from winter sleep by the spring thunder
Spring Equinox	21 March	The mid-point of the spring season; Day and night are equally long on the day
Clear and Bright	05 April	It is warm, bright, and everywhere gets green; The first day of this solar term is also a traditional Chinese festival, the Qingming Festival, also known as Tomb Sweeping Day
Grain Rain	20 April	Rainfall increases which is helpful to grain crops
Summer		
Beginning of Summer	05 May	The summer season begins
Grain Full	21 May	Grains are getting plump and not ripe yet
Grain in Ear	06 June	Wheat grows ripe marking the beginning of a busy farming season
Summer Solstice	21 June	The sun altitude arrives at the highest in the north; It has the longest daytime of the year
Minor/Lesser Heat	07 July	The hottest days have yet to come
Major/Great Heat	23 July	The hottest time of a year
Autumn		
Beginning of Autumn	07 August	The autumn season begins
The End of Heat	23 August	The hot summer is coming to an end and the heat stops
White Dew	08 September	It is getting cooler and dewdrops appear on grass and trees in the morning
Autumn Equinox	23 September	The mid-point of the autumn season; The temperature has begnn to decrease; Day and night are equally long on that day
Cold Dew	08 October	The dews are becoming frost
Frost Descent	23 October	Frost appears, and the temperature begins to descent

Part One

Winter		
Beginning of Winter	07 November	The winter season begins
Light Snow	22 November	It begins to snow but can't accumulate a lot on the ground
Heavy Snow	07 December	It begins to snow heavily
Winter Solstice	22 December	It has the shortest daytime of the year; After that day, many places in China go into the coldest period; People in northern China eat dumplings on that day and sweet balls of glutinous rice for the people in the South
Minor/Lesser cold	05 January	It is getting colder, but the coldest days are yet to come
Major/Great Cold	20 January	It is the coldest time of the year

In the Qingming festival, folk customs include avoiding lighting fires, visiting the graves and sweeping the tombs, going on family outings, playing on swings, and putting willow branches on front doors on the day.

Lighting fires to cook food is avoided, so people participate in physical activities and take exercise in case the cold food affects their health. As people go to sweep tombs, they take with them meats, desserts, fruit and others as offerings to their ancestors.

Food customs vary from region to region. In regions south of the Yangtze River, Tomb-sweeping Day is associated with the consumption of qingtuan, which are green dumplings made of glutinous rice and barley grass.

People from South Shanxi are used to steam coiling dragon-shaped buns stuffed with walnuts, jujube, and beans, with an egg in the middle of the dragon. This is called zifu, or the fortune of children. On this particular day, people make a big zifu which carries the wish for a harmonious and happy family and offer it to their ancestors. After sweeping the tombs, the whole family will share the big bun together.

Qingtuan (青团)

Zifu

Origin of the Qingming Festival

The festival originated from the Hanshi (Cold Food) Festival, which was set to remember Jie Zhitui, a nobleman of the state of Jin (modern Shanxi) during the Spring and Autumn Period. He followed his master Prince Chong'er in 655 BC during his years of exile around China. Supposedly, Jie once even cut meat from his own thigh to provide starving Chong'er with soup. Once Chong'er was enthroned as duke, however, he rewarded all those who had helped him but only passed over his loyal retainer.

Prince Chong'er and his followers including Jie Zhitui during his exile around China

With big disappointment, Jie finally retired into the forest around Mount Mian with his elderly mother. Hearing that news, Duke Wen went to the forest but could not find them. He then ordered his men to set fire to the forest to force Jie out. But Jie did not want to leave his mother behind in escaping from the fire. They both died as a result. The duke, overwhelmed by remorse, renamed the mountain as Mountain Jie, ordered his men to bury Jie there and erect a temple in his honour. He also ordered three days without fire but to eat only cold food to offer his condolence.

The people of Shanxi subsequently revered Jie as an immortal and avoided lighting fires for as long as a month in the depths of winter. The practice was later developed into 3 days around the Qingming solar term in mid-spring.

The present importance of the holiday is credited to Emperor Xuanzong of Tang. At the time, wealthy people in China were reportedly spending fortunes on too many extravagant and ostentatiously expensive ceremonies in honour of their ancestors. In 732 AD, Xuanzong sought to curb this practice by declaring that such respects could be formally paid only once a year, on Qingming.

2.3 Duanwu (Dragon Boat Festival) Festival

The Duanwu Festival (also known as the Dragon Boat Festival) occurs on the 5th day of the 5th lunar month of the traditional Chinese lunar calendar, a day when Qu Yuan (340 BC–278 BC), a great poet and one of the world four great cultural celebrities, drowned himself in the Miluo River out of his unfulfilled patriotism when his home State Chu was invaded by State of Qin during the Warring States period of the Zhou Dynasty.

Customs differ slightly from place to place, including wearing perfumed medicine bags, racing dragon boats, demonstrating martial arts, smear-

Qu Yuan

ing children with realgar, drinking realgar wine and eating preserved duck egg, zongzi (pyramid-shaped glutinous rice dumplings wrapped in leaves), and seasonal fruit.

Zongzi, realgar wine, and preserved duck eggs

Children face painted with realgar

Dragon boat race

2.4 Zhongqiu (Mid-Autumn) Festival

On the 15th day of the 8th month of the lunar calendar with a full moon at night, ethnic Chinese people celebrate the Zhongqiu Festival, literally translated as the Mid-Autumn Festival, which is also called the Festival of August, and the Reunion Festival.

Originally, the occasion was for family and friends coming together to celebrate the harvest of crops on the day when the moon is the brightest and roundest. Following the socio-cultural development of the country, the tradition gradually changed to moon worship and moon gazing with family and friends. The purpose of family reunion is very much emphasised and that is why the festival is important to Chinese people.

Chinese Food Culture

Another hallmark tradition of this festival is to display lanterns with riddles written on and guess the answers.

Mooncake is a signature of this festival. In Chinese culture, the shape of full the moon symbolizes wholeness and reunion. Thus, the making, sharing, and eating of mooncakes signifies the entirety and unity of families.

Part One

2.5 Chongyang (Double Ninth) Festival

Chongyang festival traditions include making and eating Chongyang cake, drinking chrysanthemum liquor, wearing zhuyu (Chinese cornel dogwood) leaves, climbing mountains, and admiring chrysanthemums in parks. Chongyang Cake is made of various kinds of rice flour (glutenous rice, brown rice etc.), beans flour (red beans, mung beans etc.), nuts (e.g. chestnuts), and dried fruit (e.g. jujube). The cake is steamed and sometimes can be as tall as nine layers.

Drinking Chrysanthemum liquor and wearing zhuyu leaves are two of the oldest traditions associated with Chongyang Festival. Both chrysanthemum and zhuyu are considered to have cleansing qualities and are used on other occasions to air out houses, repel insects, detoxicate, and prevent/cure illnesses. Wearing zhuyu leaves in a pouch is not as popular as before. In the old days, people believed the plant could dispel evil and keep disasters away.

Wearing zhuyu (Chinese cornel dogwood leaves)

Drinking Chrysanthemum liquor

Chinese Food Culture

Chrysanthemum display in parks

Since 1989, the day of Chongyang has been designated as Senior's Day, highlighting Chinese tradition of caring for and respecting the elderly. The new meaning of this festival encouraged new ways of celebration. For example, families with elderly members will choose to spend the day climbing mountains as a means to wish for longevity and happiness of their elderly parents.

The Origin of the Chongyang Festival
Legend says during the early Eastern Han Dynasty (25 AD - 220 AD), a pandemic befell on villages along Ru River. The devil of pandemic lingered around and claimed the life of many among them, were Hengjing's parents. Determined to expel the devil, Hengjing went through extraordinary hardship to find an immortal in Mount South East to learn from him the skills of a great swordsman.

Part One

Years later, Hengjing mastered his skills. One day, the immortal told him that the devil would appear again on the 9th day of the 9th month and he should go back to defeat the devil. The immortal gave Hengjing a bag of dogwood leaves and some chrysanthemum liquor to protect his fellow villagers from the devil. Hengjing returned to his village. In the morning of the ninth day of the ninth month, Hengjing led all the villagers to be on top of the nearest mountain. He gave each one of them a dogwood leaf and a sip of Chrysanthemum liquor. He then went to Ru River alone to fly with the devil and eventually defeated it.

Since then, the custom of climbing mountains, drinking chrysanthemum liquor and wearing a pouch of Chinese cornel dogwood leaves on the ninth day of the ninth lunar month have become a tradition.

2.6 Dongzhi (Winter Solstice) Festival

Dongzhi (Winter Solstice) is an important festival celebrated by the Chinese and ethnic Chinese in East Asia during the Dongzhi solar term (winter solstice) on or around 22 December.

Like the Qingming Festival (Tomb-sweeping Day), Winter Solstice is also one of the 24 solar terms.

Dongzhi literally means "the arrival of the coldest time in winter". As it says on the tin, it indicates that after the day, many places in China go into the coldest period in winter. Since most outdoor physical activities are limited at this time of a year, according to Chinese traditional dietary practices, people need to take more rejuvenating tonic foods to nourish the human body in light of the cold temperature.

Chinese Food Culture

Food customs for the day of the Dongzhi Festival vary in different regions of China but the tradition of having a family reunion is the same. Families in North China get together to make and eat lamb dumplings or wontons while the southerners make and eat tangyuan symbolising reunion on that day.

Tangyuan are made of glutinous rice flour and shaped in to balls which sometimes can be in bright colours. Eating glutinous rice cooked with red beans is also popular in southern China as a tradition to keep evil spirit away. Long-thread noodles are another typical Dongzhi Festival food in south China.

The Origin of Dongzhi Dumplings

In northern China, on the day of Dongzhi, people typically eat lamb dumplings or wonton soup to celebrate the arrival of the coldest time of the year. The tradition originated in the Han Dynasty and was associated with Zhongjing Zhang, a reclaimed Chinese traditional doctor. He is respected as a patron saint of Chinese traditional medical doctors in China. According to the story, Zhongjing had noticed the poor suffering from chilblains on their ears in frosty winter when he served as the Major of Changsha city. After he retired from the office, he started his medical practice in his hometown where he observed more people suffering from ear chilblains in winter. To help those poor patients, he invented "quhan jiaoer tang" (祛寒娇耳汤), or dumpling soup that expels the cold. The recipe includes lamb and herbal medicines that have a warming effect. Ingredients were made into ear-shaped dumplings and served in soup. Lamb dumpling soup was then offered to the poor for free to keep their bodies warm and keep their ears from getting chilblains in winter. From that time on, it has been a tradition to eat dumplings on the day of Dongzhi.

Part One

2.7 Laba Festival

On the eighth day of the 12th lunar month, Chinese people celebrate the Laba Festival by following the tradition of making and eating ceremonial Laba congee.

Laba congee is made of various kinds of grains, beans, dried fruit, seeds. Although the ingredients may be different from region to region, the most common recipe contains rice, jujube, red beans, longyan, almonds, peanuts, sunflower seeds, lotus seeds etc.

In some regions, Laba food customs are closely linked with Buddhism. It is believed that the Buddha had obtained enlightenment on the day of Laba. In remembering this significant moment of Buddhism and the path to enlightenment, Buddhists cook Laba congee with 18 ingredients which represent the 18 arhats.

Section 3 Food Customs in Life's Occasions

Shakespeare said that "Life is a journey from cradle to coffin". There are many milestones throughout a person's life among which the most important ones to Chinese are birth, adulthood, marriage, and death. Ceremonial rituals were developed to celebrate and commemorate these moments. Four most important and commonly practiced celebrations of life events are the arrival of a new baby, wedding, longevity and funeral. The food customs in the above-mentioned celebrations will be introduced in this section.

Chinese Food Culture

3.1 Celebration of Birth

The arrival of a new life is a great gift to the entire family. There are a few traditions to follow in the celebration of birth.

When the baby arrives, the family will announce the happy news among its relatives and friends which is called "Bao Shi", (报喜, literally means to report the good news). People who have received this happy news will send "red eggs "to congratulate the family. In some regions, two bottles of Chinese rice liquor will be given together with the red eggs as an essential gift pack. Depending on the relationship with the family, people will choose appropriate gifts to complete their gift pack. The more the merrier. The family will thank its relatives and friends by giving back red eggs and congee of Joy.

The family will arrange banquets to entertain relatives and friends to mark the first month, the 100th day, and the 1st birthday of the baby.

In Southern China, when a baby girl is born, her parents will brew alcohol for her, bury it underground and keep it until her wedding day.

3.2 Traditional Chinese Wedding Celebration

A wedding has long been regarded as a big occasion in life across the world.

Traditional Chinese wedding rituals originated thousands of years ago. Traditions have been simplified over time. Essential food customs are still an important part of a Chinese wedding. Practices vary from region to region. The customs described below are more likely to be found in Northern China.

On the wedding day, the groom collects his bride from her parents' house. When he arrives, he must overcome obstacles in the form of playful teasing by the bride's family and bridesmaids before he can take away the bride. Meanwhile, the bride serves her parents a cup of tea (in some places is a cup of Chinese rice liquor) to show her gratitude for their love and care and to express her reluctance to part.

Part One

The bride then made her way to the groom house accompanied by the groom and a few of her family members. After arriving at their bridal chamber, the couple will sit on the wedding bed drinking wedlock wine with their arms linked, a symbol of marriage. A family member of the bride will feed her a half-cooked dumpling and ask her "Is it raw (pronounced as Sheng in Chinese)?" and she must answer "Yes". In the Chinese language, Sheng (生) has double meaning. It could mean raw or giving birth. On wedding occasions, this tradition implies fertility. For the same reason, the wedding bed will be decorated with jujube (zao, 枣), chestnut (li'zi, 栗子), peanut (hua'sheng, 花生) for all of them are symbols of fertility.

The reception banquet given by the groom's family on the day of the wedding is the highlight of the wedding ceremony. Guests sit around 10-seated round dinner tables and will be treated to an eight, nine or ten course meal. It is seen as a public recognition of the union. Most of the foods commonly served have good meaning, such as a whole fish dish is served because the word for fish sounds like word for plenty, suggesting a wish for abundance. Sweet lotus seeds for dessert indicates a wish for many children.

On the following morning of the wedding day, the newlyweds will serve the groom's parents a cup of tea to show their respect. Parents' act of accepting tea and giving red envelopes filled with money is the symbol of welcoming the bride to the family.

3.3 Celebration of Longevity Party

Longevity celebration is a type of birthday celebration, usually celebrated every ten years after a person reaches 50 years of age.

Traditionally, children will prepare a birthday celebration to invite extended family members and friends on or around their parents' birthday.

Days before the birthday, children often cook steamed peaches made of rice powder or flour, and deliver them to relatives and friends, informing them the date of the celebration. On the day of celebration, appetisers usually include eggs, tea cakes, and long noodles which are commonly known as noodles of longevity. The number of courses served at the feast is thought to be as many as possible as it is a sign of fortune and longevity. In recent years, more people have opted for to host a formal meal in a restaurant.

3.4 Funerary

Depending on the age, the marital and social statues of the deceased and the cause of death, different funerary rituals are applied in different regions of China. If a family lost an elderly family member who was over 70 years old and died of a natural cause without much suffering, the family often organises a feast after the burial to thank everyone who attended the funerary. According to Chinese folk religions, this type of funerary is regarded as a white happy occasion. The feast normally comprises 9 courses, all of which are vegetarian.

Chapter 5 Customs and Etiquette in Chinese Dining

Every nation has its unique dining customs and etiquette. In most traditional Chinese dining, dishes are shared communally. Square tables are used for small groups of people. 10-seated round tables are preferred for large groups for easy sharing, particularly in restaurants. Sitting around the table and sharing dishes is a Chinese-specific dining culture embodying Confucius belief of "Harmony is most precious".

Section 1 Common Forms of Chinese Dinners

Considering the purpose of dining, Chinese feasts fall into three categories.
- Banquets that are held for courtesy. For example, the banquet that is organised to welcome important guests and diplomatic visits or to celebrate important festival occasions.
- The second type is casual meals to socialise with friends to develop and maintain friendships.
- The last type is business meals.

1.1 Banquet

A banquet is a pre-planned dining activity held for social interaction. The Chinese banquet refers to the banquet using Chinese tableware, offering Chinese food with Chinese-style table arrangements and services, and reflecting Chinese eating customs. Based on the differences in its nature and the seniority of the guests, there are state banquet, formal banquet, casual banquet, and family banquet.

State Banquet

A state banquet is a formal welcome dinner hosted by the head of state for guests attending big domestic or international events. State banquets are usually considered as banquets with the highest level of reception and the strictest protocol.

Currently, China's state banquets, depending on the purpose, can be categorised into four types:
- **Celebration Banquet** is a reception held by the head of state or government to celebrate National Day.
- **Welcome Dinner** is a formal welcome dinner hosted by the head of state or government in honour of visiting heads of state or government. Those accompany officials and ambassadors from visiting countries will attend.
- **State Reception** mainly refers to the banquet hosted by heads of state or government for major international or domestic events.

- **Tea Party** is a banquet held by heads of state or government on New Year's Eve to greet the coming year. Prominent figures from all sectors are invited to this celebration. Tea and other soft beverages, desserts, snacks, and fruit are served. Guests would also have the opportunity to watch a gala performance which creates a casual and joyous atmosphere.

Formal Banquet

A formal banquet is a high-level, sumptuous, solemn banquet with elaborate and strict protocol. It is often carefully arranged for a specific person or a group of people and held in superior restaurants or some other designated venues. It has strict requirements for the seating, the number of dishes, musical performance, speeches from the host and the guest.

Informal Banquet

An informal banquet is a simple and casual banquet for everyday friendly communications and contacts. It has less formality, emphasises not the magnitude or level, but on interpersonal relationships. In general, only relevant people are invited, not their spouses. Little stress is put on dressing, seat arrangement, the number of courses. And there is no music performances or addresses from the host and the guest.

Family Banquet

A family banquet is a feast for family members, or an informal banquet for guests at home in the host's own name. It is often held in happy occasions such as weddings, birthdays, birth, or traditional festivals. Some common ones are wedding banquets, birthday parties for the elderly, New Year Eve's dinner, and feast in the Mid-Autumn festival.

There is no special protocol requirement nor menu. the style of dish and number of courses to be served are at the discretion of the host family. In order to make the guests feel valued, welcomed and at ease in the home, host and hostess usually entertain the guests by cooking and serving to create a family atmosphere.

To take part in a family banquet, guests must be well-dressed and arrive on time out of respect to the host family as well as to themselves. When entering the house or the banquet hall, guests should first say hello to the host. For other guests, whether familiar or not, should be greeted with a nodding smile or shake of hands; when seated, if there are senior people around, guests should stand up and offer their seats to the elderly.

1.2 Informal Meals

Informal meals are very casual. The dining locations are often different, and the etiquette is minimal, there is no need to over emphasise other etiquettes.

1.3 Business Meals

A business meal is a meal for business partners to contact, keep in touch, exchange information or negotiate their deals. It is generally a small-scaled lunch. The host doesn't need to send a formal invitation, and the guests don't need to respond formally in advance. The time and place can be decided at any time.

1.4 Buffet

Buffet is a modern way of dining introduced from the west in recent years. It needs no seating arrangements, nor a written menu. It serves starch food, dishes, and beverages which are placed in a public area where the diners generally serve themselves. There are not many requirements over dining etiquette. It is convenient for both the host and the guests. After finishing eating, dinners are expected to send the tableware to the designated area or to leave them neat and clean on the table for the waiting staff to collect.

Part One

Section 2 Chinese Banquet Arrangements

2.1 Definition

In a broad sense, Chinese banquet arrangements refer to a range of plans associated with selection of food and beverage, service order and activities at a banquet.

When arranging a Chinese banquet, not only the choice of dishes and food displays need to be considered but also overall dining environment, venue set-up, equipment and facilities, food production processes and services, as well as the effective implementation of the above-mentioned plans.

In a strict sense, banquet arrangements only refer to the types of food served at a banquet, proportion and service order of each type. The knowledge of a banquet arrangement will help us to understand the sequence of making and serving dishes at banquet. In this unit, the definition of a banquet arrangements should be understood in the strict sense, unless stated otherwise.

Food served at a Chinese banquet contains cold dishes, hot dishes, starch dishes (also known as banquet snacks), and fruits.

Normally a range of alcoholic and non-alcoholic beverages will be provided at a Chinese banquet for guests to choose from based on their preference. Since selection of beverages is not part of the Chinese banquet arrangement, they will not be discussed in this section.

2.2 The Common Banquet Arrangements

The types of food served in a banquet and their proportions are varied in different forms of banquet. However, in terms of the service, there are only two main styles, the table service style and the buffet style. To enable a straightforward introduction, the banquet service discussed in this section will be limited to table service style.

After a period of rapid reform in recent years, the Chinese banquet has adopted a 4-phased serving procedure (as shown in Figure 5-1) which is commonly used in all regional cuisines.

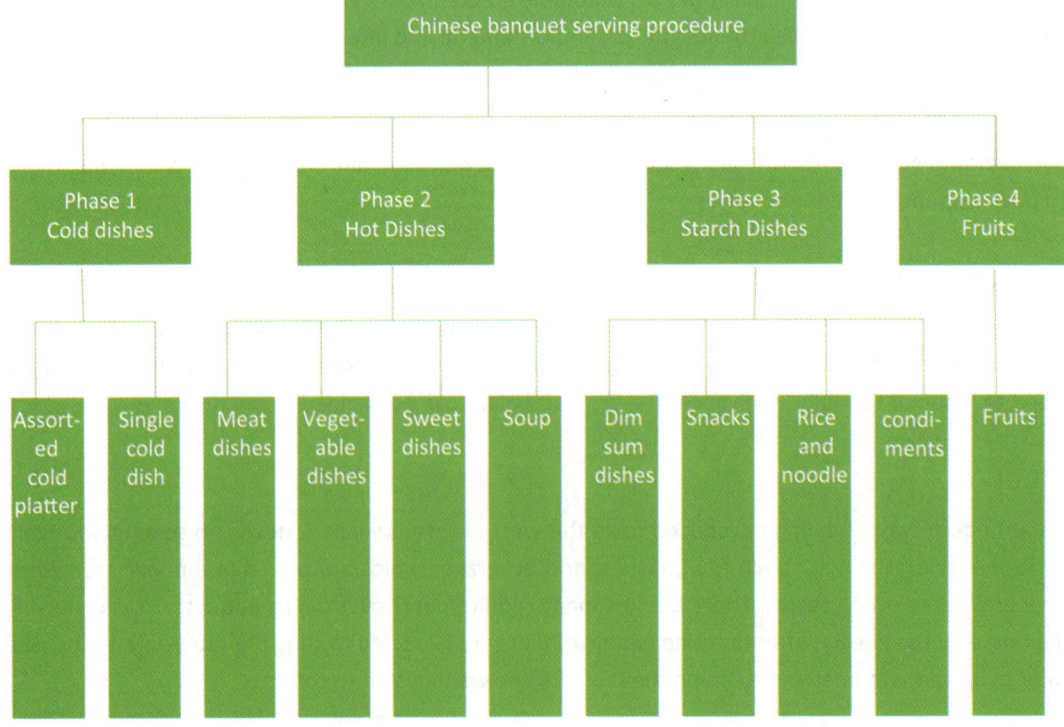

Figure 5-1 4-phased Chinese banquet serving procedure

Having maintained the advantages of traditional Chinese feasts, this serving plan is now widely adopted and accepted by the Chinese people. However, it only provides guidance on banquet menu design, changes can be made to suit specific situations.

Although a Chinese banquet usually follows these 4 phases, serving order of dishes within one phase may differ from one region to another. For example, Cantonese people tend to consume soup prior to other hot dishes. Therefore, for Cantonese cuisine, the soup dish is often the first hot dish served after the cold dishes, while for the Sichuan, Shandong, and Huaiyang cuisines, the soup is more often served as the last hot dish.

2.3 Common Dish Combination in Chinese Banquet

Cold Dishes

The arrangement of cold dishes, or cold appetizers has many varieties. Common forms include the following:

AN ASSORTED PLATTER

A COMBINATION OF 6 TO 8 TYPES OF SINGLE, MAIN-INGREDIENT COLD DISHES

Part One

COLD DISHES SERVED ON INDIVIDUAL PLATES FOR EACH GUEST (AN ADAPTATION OF CANAPÉS IN WESTERN CUISINE)

The ratio of meat and vegetables can be 4: 1, 3: 1 or 2: 1. Cold dishes account for 10% to 20% of the cost of the entire banquet dishes.

Hot Dishes

Hot dishes usually include stir-fried dishes and elaborate dishes. Stir-fried dishes require smaller-sized ingredients and less cooking time and are usually served in plates of 8" to 10". The number of stir-fried dishes served in a banquet should be limited to 2 to 4 courses. When designing the menu for a large-scale banquet, courses of stir-fried dishes should be kept to the minimum or avoided entirely.

Elaborate dishes are normally made of larger-sized main ingredients (such as large chunks of meat, a whole chicken/duck/fish/suckling pig, or half a rack of ribs, etc.), and served in 12" to 14" plates. There may also be smaller-sized ingredients, but the proportion is less than that of large ones. There are often 6 to 10 courses of elaborate dishes.

In a traditional banquet, the first course is of significance. It represents the class of the banquet, and thus is mostly made of superior ingredients. It is followed by some roast dishes, such as roast port or roast duck served with wrappers and two soup dishes. This arrangement is still common in modern style banquets.

Part One

Vegetable dishes take up around 20% -30% of the total hot dishes. The cost of hot dishes accounts for 70% - 80% of the total banquet cost. Sweet tasting dishes are a distinctive type of food in a banquet menu. Typically, there will be one or two courses of sweet dishes served in between meat dishes and vegetable dishes. One of them is dry cooked and the other one is usually cooked with juice.

Starch Dishes

Starch Dishes are staple food. Within this category, Dim Sum (点心) takes a small proportion (2-4 courses) on a banquet menu unless it is a Dim Sum feast. Other staple food options include a variety of rice and noodles dishes. For example, Yangzhou fried rice, rice in seafood soup, and noodles in chicken soup to name just a few. Staple food accounts for 5% to 10% of the total cost of the banquet.

Fruits

Fruits are usually prepared and served on a platter for guests to share. Occasionally, 1 or 2 kinds of fruit may be served directly on the table. Fruits account for 1% to 2% of the total banquet cost.

Section 3 Chinese Banquet Etiquette

3.1 Seating Arrangement and Table Layout of Chinese Banquet

Seating Arrangements

Generally, Chinese banquet tables are round. The seat directly facing the entrance is reserved for the host or the most important or highest ranked attendee. In a formal setting, the closer that people are seated to the host, the higher their rank is, with right taking precedence over left when distance is equal, as in Chinese culture, right is considered superior to left.

In a banquet with more than one table, there should be a representative of the host at each table to serve as the host for the table. Their seats can either face the same direction as the host at the main table or face the host at the main table. The number of people at each table should be an even number which should not exceed 10.

Figure 5.2 is samples of two different seating arrangements in a formal banquet involving international guests. In Plan A, while the host faces the entrance, the guest of honour sits to the right of the host and the second guest of honour sits to the left of the host. The second host sits on the opposite side of the host, whose right-hand side is the third important guest and the left-hand side is the fourth important guest.

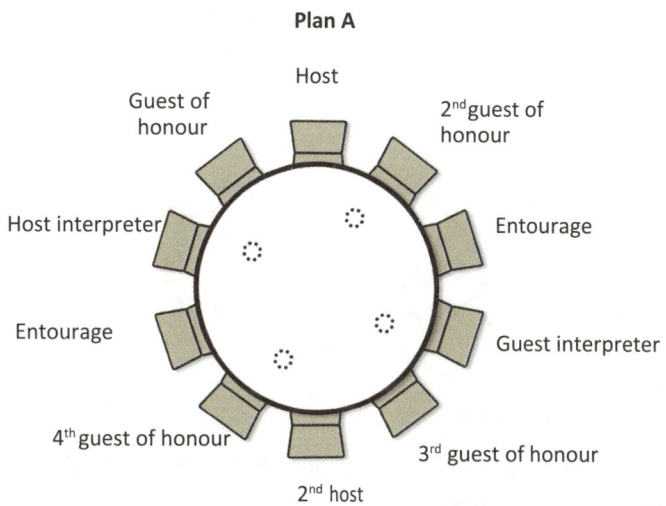

Part One

In some cases, Plan B seating plan is applied where the third important guest sits to the left of the host and the 2nd guest of honour sits to the left of the 2nd host. Other seats are reserved for interpreters and entourages.

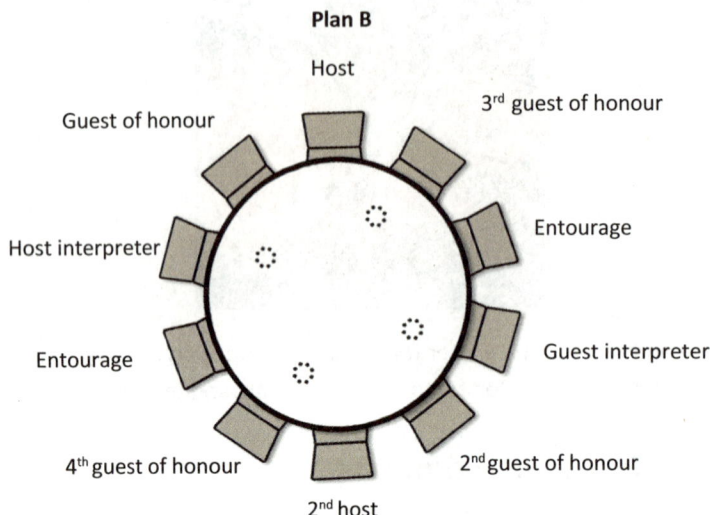

Seats will be pre-arranged for all guests in formal state banquets. However, in most large and medium sized fewer formal banquets, only the seats of the main table will be arranged beforehand. Other guests will find an allocated table number in their invitation. Upon their arrival at the banquet venue, guests will be led to their table by concierge where guests have the freedom to select their seats.

Table Layout

In banquets with two or more tables, the main table would be placed in the most prominent position, either at the centre or on the top. Therefore, table layout can take a wide range of shapes such as triangle, square, trapezium, circle, diamond, H-shape as shown in Figure 5.3. Similar to the arrangement of seats, it is a Chinese tradition that the closer that tables are to the main table, the higher the rank of guests is, with right taking precedence over left when distance is equal.

Figure 5.3 Table layout patterns in a Chinese banquet

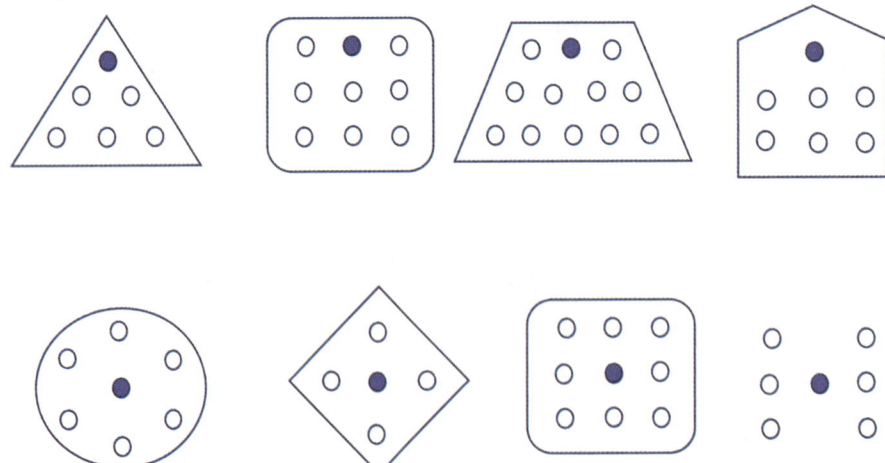

3.2 The Placement and Use of Tableware

As an important part of dining customs and banquet etiquette of Chinese people, a proper placement of Chinese tableware is required in a Chinese banquet.

84

Chinese Food Culture

The Placement of Tableware in Chinese Banquet

Figure 5.4 A Proper Place-setting in a Chinese Banquet
- The positioning of food plate (2). A food plate, also known as bone plate or eating plate, is the main plate used by guests to hold food. It serves as the central plate at each place. The food plate is placed on top of a decorating plate (1). The logo or design pattern on the decorating plate should be facing the guests.
- Placement of saucer (3), soup bowl and spoon (4&5). Saucer is placed 1cm apart from the food plate with the centres aligns. Soup bowl and spoon are placed to the left of the saucer, 1cm apart, with the centres align. The handle of the spoon pointed to the left.
- Placement of chopstick rest (6), serving spoon (7), chopsticks (8), serving chopsticks (9), and a small pack of toothpicks (10). The chopstich rest is placed to the right of the saucer and is in line with the centre of the saucer. The end of chopsticks should be 1.5cm away from the edge of the table. The handle of the serving spoon and food plate are 3cm apart.
- Placement of glassware. Wine glass (13) is placed 2cm apart and above the saucer with the centres align. Liquor glass (14) is placed to the lower right of the wine glass while water glass/water goblet (12) is positioned to the higher left of the wine glass. Three glasses are in line forming a 30-degree angle from the horizon with a 1.5cm distance between each other.
- Placement of menu and other communal utensils. In a high-profile formal banquet, each guest will have a menu in front of them or at least four copies of the menu will be provided per table. In an ordinary banquet, one or two copies of the menu will be placed on each table. Ashtrays are not provided in non-smoking restaurants.
- Display of table napkins. The Napkin arrangement is designed to highlight the banquet theme. Therefore, it needs to consider the size and theme of the banquet. Napkins are usually placed in the water glass or on the food plate.

Part One

Table-setting Example A

Table-setting Example B

The Use of Chinese Tableware and Taboos

Chopsticks

Chinese banquet table manner mainly refers to "chopsticks manner". The picture below explains how to use chopsticks properly. The position of the thumb is roughly 2/3 of the full length from the tips. By doing so, it not only looks elegant but also gives more control over the use of chopsticks. Holding the chopsticks either too high or too low will be considered inappropriate.

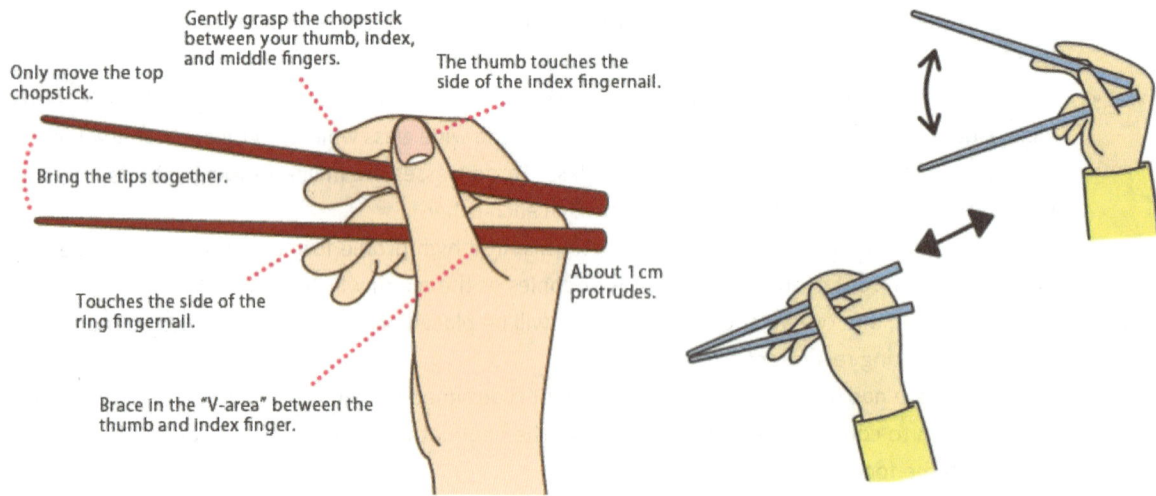

Figure 5.5 How to hold Chopsticks Properly

Chinese Food Culture

Chopsticks Manners (Taboos)

To use chopsticks to reach out to dishes that are more than one arm's length away or to cross over other guest's chopsticks to fetch food

To place chopsticks crossed on a bowl or a plate. This is reminiscent of funerals

To tap at an empty bowl with chopsticks before a meal

Using chopsticks to "search" for stuff in a shared dish

Standing the chopsticks upright in rice as its is considered to resemble the ritual of incense-burning to symbolise feeding the dead or death in general

Licking the chopsticks

87

Part One

Using chopsticks that have food stuck on them

Using chopsticks to fetch food from the same dish for more than three times at a one go

Holding chopsticks while hovering over various dishes

Using chopsticks to pick teeth

These taboos are mostly related to hygiene and courtesy.

Serving Spoon

The serving spoon is mainly used to scoop food from shared dishes, or sometimes to assist the chopsticks in fetching food. Try to avoid using the spoon alone to fetch food from dishes. When spoon is used to get food, a small amount should be taken each time to avoid spillage. A suggested way to use spoon when picking up juicy food is to rest the spoon on top of the dish for a short while until there is no liquid dropping down, and then move it back to one's own plate.

When the spoon is not in use, it should be left on the food plate. In the case of accidentally picking up more food than you intended from a shared dish, it is not recommended to pour food back into the dish but to place it on your food plate and finish it. Do not stir hot food to cool it down with a spoon nor blow on food with mouth. Hot food should be left on the food plate to cool down. Do not put the spoon in your mouth, or repeatedly suck and lick food on it.

Food plate

The food plate is mainly used to hold food that is picked up from shared dishes. It is recommended neither to take too much food nor too many kinds of dishes to leave on the food plate at one time. Mixing various dishes on the food plate spoils the look and the taste of food. Bones and other unwanted bits should not be spitted to the table or floor, instead, they should be taken out of the mouth and left on the top side of the food plate.

Water Glass

A water glass is used to hold soft drinks such as water, sparkling drinks, fruit juice. It is not for wine or spirit and should not be placed upside-down.

Toothpick

When picking teeth in public, cover the mouth with one hand or napkins. Do not keep the toothpick in the mouth for too long. Foremostly, toothpicks should not be used to stick in food and pick food up.

Bowl

A bowl is for rice and soup. When eating from a bowl, one should hold the bowl with the left hand, and hold chopsticks or a spoon with the right hand.

Part One

Others

Before the meal, a wet facial towel will be provided to each for cleaning their hands. After use, leave the towel on the side plate to be taken away by the waiter/waitress. Sometimes, at the end of the banquet, another round of wet towel distribution will be offered. Different from the first round of towels, the towel is only used to clean the mouth.

3.3 Dining Etiquette in a Chinese Banquet

Dining etiquette is very much emphasised in banquets in all aspects from dress code to manners. Dresses in cheerful and bright colours are for weddings while black and plain-coloured clothes are most suitable for attending funerals. When attending a banquet of any kind, punctuality is the key. Leaving early or staying for a very short period of time are considered to be rude and hostile to the hosts. For genuine reasons that you plan to leave in the middle of the banquet, inform the host before the banquet starts. And if possible, leave the table after the main courses (in phase 2) are served. When the situation emerges that you must take an unplanned early leave, explain to your host the reason and ask for forgiveness, and then express your gratitude to the host's hospitality.

Chinese people believe that being a guest should avoid causing too much inconvenience to the host. Therefore, it is important to agree with the host's arrangement. When you sit next to elderly guests, you are expected to provide assistance to settle them in. Showing respect to the host and others by paying attention to programmes and activities arranged for the banquet is a virtue.

Toast is a very important part of a Chinese banquet. In a small-sized banquet, the host will propose toasts to all the guests one by one regardless their ranks, starting with clinking glasses with the guest of honour and then the rest. If there are many guests present at the banquet, clinking glasses can be replaced by simply lifting glasses for a toast. While the host and the guest of honour are giving a speech, the rest are expected to stop eating or toasting and listen to them carefully. Guests also can toast and clink glasses with each other to extend goodwill and to heat up the atmosphere. When receiving a toast, you should respond positively and always propose a toast back. Consumption of alcohol at a banquet needs to be mindful to prevent getting drunk.

Attending a banquet, one always needs to maintain good manners. When taking food from shared dishes, small amount is recommended at each time. It is considered inappropriate to stand up to reach food. If you are served with a dish that is not so much to your liking, it is better to take a small amount than reject it at all. Talking and laughing with your mouth full could potentially spray food all over the table, therefore they are considered very rude. Chewing food or drinking soup quietly with mouth closed is a widely accepted norm. If the soup is too hot, wait until it cools down. Fish bones should be placed on the food plate. Picking teeth or nose or digging ears at a banquet is inappropriate.

Section 4 Tea and Wine Culture in Chinese Banquets

4.1 Wine Drinking Etiquette in Chinese Banquets

Selection and Consumption of Chinese Wine

In a Chinese banquet, Baijiu (white wine), Huangjiu (yellow wine), Yaojiu (medicinal liquor), and Pijiu (beer) are commonly available. Baijiu is the most common liquor in China; it can be taken alone, or with dishes. Drinking Baijiu together with other alcoholic beverages would make people feel drunk quite easily.

Drinking strong alcoholic beverages such as Baijiu, Huangjiu and Youjiu, Chinese liquor classes of normally used. It is made of ceramic or glass and is small (similar to a shot glass). Baijiu (white wine) is served at room temperature with no ice cube or water added; while Huangjiu (yellow wine) can be warmed up before drinking. Especially in winter, many regions in China maintain the tradition of drinking warm Huangjiu.

Part One

How to Pour Chinese Wine

When a guest is served by a waiter/waitress, the guest does not have to lift the liquor glass but should say thank you. If it is the host who pours the wine for guests, guests must hold the glass up or even stand up and bow slightly to show their gratitude. Guests can also use the "finger bowing etiquette", which is, putting the index finger and middle finger of the right hand (even if you left-handed) together to knock the table gently three times.

When the host pours wine for guests, the general rule is to treat every guest equally, at the same time, be mindful of the amount of wine to be poured. Normally, Baijiu glass and beer glass can be filled up to 80%.

Things to Remember About Proposing a Toast

A toast is a ritual to express honour or goodwill. In a formal banquet, it will be the host to initiate a toast for something or somebody so honoured. A toast normally involves a short speech accompanied by a drink. Sometimes, the host and the guest of honour will deliver a specially prepared message which should be as short as possible.

Depending on the type of banquets, a guest can give toast to his/her fellow guests by approaching someone to clink glasses first and then drink together or raising your glass towards someone from a distance and drink. In both cases, drinking is an indication of expressing goodwill, therefore it is essentially toast. More elaborate and formal toast normally occurs after guests are settled in and before the meal starts.

In China, the ritual of toast includes the following steps. Firstly, when a toast is proposed, the guest whom the toast is for should stand up and hold the glass with both hands, look at the person, express goodwill back, raise the glass to eye level or clink the glasses, say "gan bei" (bottoms up) to the person, take a sip from the glass or finish the whole glass, finally look at the person again and smile. The ritual applies to people who do not drink except those who need to simply use the glass to touch their mouth without taking the drink.

When clinking the glasses, make sure your glass is lower than those who are senior than you either because of their age, social status, or rank to show respect.

When proposing a toast to everyone present at the banquet in turn, make sure to start with the eldest, or the one with the highest rank, or the host.

For any reason that you could not drink alcoholic beverages, you could either entrust a relative, a friend, or a college to respond to toast on your behalf or you could drink tea instead. The person who proposed that toast should not force others to drink as a courtesy.

4.2 Tea Etiquette in Chinese Banquets

In China, brewing and serving tea is an essential etiquette when receiving guests both at home and in a banquet.

Before the banquet begins, the host will show his hospitality by serving tea according to the guests' preference. Commonly available tea choices include green tea, black tea, oolong tea, jasmine tea, and dark tea. Different tea ware will be used according to the type of tea. The majority of tea ware is made of pottery or ceramic. Normally, lighter tasted tea will be provided before the meal unless it is specifically asked by the guests. Tea will be served hot. As a general rule, the teacup is normally filled up to 70%.

Part One

When serving tea to others, use both hands to show respect. If there are many guests to serve, always follow the principle of guests first and then the host, the guest of honour first and then the rest of the guests, lady first and then gentlemen, guests first and then the younger guests. Traditionally, Chinese people will avoid pointing the mouth of the teapot to the guests. Other simpler serving orders include going clockwise starting from the guest of honour or on a first come first served basis.

Before the banquet finishes, another round of hot tea will show the hospitality of the host.

Chapter 6 Outline of the Regional Culture, Food History & Culinary Style, Tianjin

Section 1 An Overview of Tianjin Food Culture

Tianjin is located on the plains near the Yellow River, giving it rich soil, appropriate latitude and sufficient sunlight, thus, its environment is favourable for agricultural development. Originated from the Central Southern-plains, Tianjin Xiaozhan rice and Baodi garlic are famous around China and overseas. Numerous mountains in the Northern areas breed different kinds of nuts and fruits, like chestnuts, walnuts, hawthorn, red fruit and persimmon, as well as high-quality grapes to make wine. On the Eastern side, there are two big ports Da Gu and Bei Tang. The wide waters resulted in a great variety of seafood products, including lake fishs, sea fishs, shrimps, crabs and mussels. The high yield of seafood and the high quality caused the custom of eating seafood in Tianjin. Not only do Tianjin people love seafood, they eat according to the Climate. As Tianjin is a place that gathered cultures from "5 directions", food culture in Tianjin appreciates high-quality and fresh food, as well as luxurious style cuisine.

Section 2 Tianjin Local Food Culture

The tradition of worshiping ancestors and Chinese folk religion gods and the rituals associated with the celebrations of birthday, wedding, and new year have existed for a long time in Tianjin. As time passed by, following the continuous development of Chinese civilisation, some traditions were reformed. Noticeably food customs gradually replaced ritualistic formalities whilst food fair replaced religious gathering. Eventually the culture of making and consuming delicious food as a means of celebration emerged.

Chinese New Year is the most important festival of the year. Like elsewhere in China, making and eating dumplings on the Chinese New Year's Eve are a central theme in Tianjin. The filling of dumplings served on this occasion normally contain at least three or four ingredients. Traditionally, the first day of Lunar New Year was a vegetarian day in Tianjin to wish for a year without strategic incidents. The second day was to worship the Fortune God. People would make a noodle dish with as many ingredients as possible symbolising wealth. Gradually, the tradition started to change. Nowadays, people would have vegetarian dumplings on the first day of Chinese New Year, noodles on the second day, and Hertze (盒子, a form of dumpling, a symbol of bringing the fortune home) on the third day. On the 15th day of Chinese New Year, people in Tianjin celebrate Lantern Festival by eating sticky rice balls with sweet fillings, representing the union of all family members. On 2nd February in lunar calendar, it is the Dragon Head-raising Day when dragon awakes from its hibernation and raises its head. According to the Chinese folk legend, the dragon is an auspicious animal that controls clouds and rains. After this day, spring is coming and there will be more and more rains. Being part of North China region, Tianjin has the tradition to eat Mentze, a staple food made of sweet potato or Chinese yum on the Dragon Head-raising

Day.

Early spring would be the season for harvesting huang shrimps (晃虾). Huang (晃) means in the blink of an eye. As suggested by its Chinese name, the supply of huang shrimps only last for a short period, which makes it a sacred seafood product. In March, there would be crabs. Around the Qing Ming Festival, yellow croaker and king prawns would appear on the market followed by noodle fish (Pacific sandlance) and white rice shrimps in May and June to make the traditional Tianjin dish "Noodle fish omelette".

On the day of the "Beginning of Autumn", people have the tradition to "bite the autumn" by eating watermelons. It was suggested that this could prevent diarrhoea. To celebrate the Mid-Autumn Festival, in addition to the festive mooncake, people would also eat folkish delights such as autumn crabs, homemade flatbread, and congee.

In winter, Tianjin people prefer hearty food, hence the custom of eating wonton soup on Winter Solstice. Hotpot is an all-time favourite winter hearty food for Tianjin people. "Lamp hotpot with clear soup", "Sanxian hotpot" and "Assorted hotpot" are the most popular options.

Section 3 Tianjin Local Cuisine

3.1 The Formation and Development of Tianjin Local Cuisine

Figure 5 The River Hai of Tianjin

Tianjin is located at the north-eastern extremity of the North China Plain, with the Bohai Sea on its East and the Yan Mountains on its North. Rich natural resources helped Tianjin to develop an important role in Chinese cuisine.

The construction of the Grand Canal (Da Yunhe) beginning in 605 A.D. during the Sui dynasty (581 A.D. - 618 A.D.) created series of waterways in eastern and northern China that link Hangzhou in Zhejiang province with Beijing and made the land where Tianjin is located today a leading river transport centre.

The rise of Tianjin local cuisine traces back more than 600 years to the Ming Dynasty when the development of Tianjin as a city began. The settlement became a garrison town called "Tianjinwei" (Defence of the Heavenly Fort) in 1404 under the ruling of Emperor Cheng Zu (1360-1424) to protect Beijing, the capital city . A large military based was built. Troops were sent from all over the country to station here. The soldiers brought their hometown food to Tianjin. Gradually, a new style of cuisine, which was largely influenced by Shandong (Lu) cuisine, started to appear. It was regarded as the early form of Tianjin Cuisine. By the middle of the Qing Dynasty,

Tianjin had become a commercially prosperous city due to the success of being a river transport and salt trading centre. Tianjin local food customs were inevitably influenced by business travellers and immigrates from other regions of China, such as Shandong, Shanxi, Henan, Jiangsu, and Zhejiang.

After the Xinhai Revolution in 1911, many abdicated royal family members and retired high-rank officials from the previous regime chose to reside in Tianjin to continue their luxurious lifestyle. The aristocracy and politicians together with wealthy merchants, retired warlords formed up a special group of customers who had a desire for high quality food. Their food related stories were spread out through the writing of literati. The food trends led by them became a catalyst of the further development of Tianjin Cuisine. it was a time when many food customs and cooking techniques of other regions were introduced to Tianjin, a lot of tales were told, and a great deal of newly created dishes became the trends.

With over 400 years of development, Tianjin Cuisine has become popular nationally and internationally. Following the socioeconomical development in China, people have changed their perception of food and diet. The desire for quality food and good food experiences were no long a privilege of the high society. Common people's food and drink consumption for leisure becomes the main market of food industry. Customers nowadays like new trends and nature products with a strong brand and culture awareness. With the help of Tianjin's rich resources, Tianjin style cuisine has gradually developed into a complete cuisine system, an integration of Han cuisine, Chinese Islamic cuisine, temple cuisine, folk cuisine and folkish delights.

3.2 Characteristics of Tianjin Cuisine

Tianjin Cuisine has five distinctive characteristics.

3.2.1 Specialised in Cooking River Fish and Shrimp and Seafood Ingredients

Due to its location, Tianjin is rich in "river fish and shrimp" and "seafood" which are two major sources of the umami taste in Chinese cuisine. Hence, they are often referred to as the "two umamis". Tianjin people love the "two umamis". Consequently, Tianjin chefs have developed exquisite techniques to cook them. Taking fish for an example, based on the flavour, texture, size, and other characterises of the type of fish, Tianjin chefs would be able to use techniques such as stewing, braising, quick stir fry, and boiling to create delicious and sophisticated fish dishes. There are even more techniques for cooking prawns, such as "pan-fried prawns", "prawns made into two dishes", and "prawns made into three dishes". The latter two involve cooking different parts of prawns using different methods so that two or three dishes will be produced at the same time.

3.2.2 Eating According to Season

The custom of eating the right food in the right season has existed for a long time in Tianjin. Tianjin people are strict in keeping their food tradition to eat according to season. The restaurants in Tianjin are good at launching seasonal dishes for customers.

3.2.3 Excellent Seasoning Techniques

If we consider the harmonious blending of the five tastes of salt, sour, sweet, bitter, and pungent is the cornerstone of Chinese cooking, the flavour of a dish that is created through this cooking process would be the soul of the dish. Tianjin chefs have accumulated a great deal of experiences in creating flavours from years of practice. In addition to apply common seasonings such as salt, cooking wine, and soy sauce, chefs would also use pepper oil, ginger puree, and red caramel. Also, adding soup stock to enhance the flavour of a dish is a specialty of Tianjin Cuisine. Hence, there is a particular emphasis on soup making techniques. Clear soup, white soup, and vegetarian soup are the three main types of soup in Tianjin Cuisine.

3.2.4 Unique Cooking Techniques

Tianjin Cuisine involves using all major cooking techniques of Chinese Culinary Arts and is specialised in the methods of Peng, Chao, Shāo, Bao, Liu, and Pa. In particular, the Pa technique is unique in Tianjin Cuisine. It requires laying all the ingredients in a wok in a designed pattern with the intended top side facing down, regardless the ingredient is an entire duck or chicken, or finely shredded meat. The pattern must not to be disturbed during the cooking process so that the finished dish will be in its expected shape.

3.2.5 Great Variety

The great variety of Tianjin Cuisine can be observed in the following ways.

THE VARIETY OF FLAVOURS

Tianjin Cuisine chefs are highly creative in terms of creating flavours. They are able to consider the various preferences of an ethnically diverse population of Tianjin, many were relocated from other regions of China, to create compound flavours to satisfy the taste buds of all kindss of customers.

THE VARITY OF DISHES

Tianjin Cuisine includes Han Chinese dishes, Muslim dishes, vegetarian dishes, and local folkish delights.
Muslim dishes make up an important part of Tianjin Cuisine. Locally produced ingredients such as river fish and shrimps, seafood, beef, mutton, and duck are cooked by using multiple cooking techniques.
A wide range of **vegetarian dishes** can be found in Tianjin Cuisine. They are cooked in meticulous techniques with superb skills and are as tasty as non-vegetarian dishes. There are more than 100 types of vegetarian dishes that are suitable for casual gathering or fine dining.
Tianjin local folkish delights are dishes created by common people in rural areas of Tianjin. For example, "8 seafood"- the wild-caught seafood freshly cooked in seawater by fishermen in Hantang area, "drunken crab from the River Ning" from River Ning area, "braised fish" from Jinghai area, "thousand-layered flatbread with meat" from County Ji, "Yong Yang food box" from Wu Qing, and fried mung bean pancakes from Bao Di.

Section 4 Representative Dishes of Tianjin Cuisine

4.1 The Best Restaurants and Their Notable Dishes

Bao Xuan Seafood Restaurant
Bao Xuan Seafood Restaurant is a restaurant chain with a distinctive and clear brand identity – to provide dishes that use local river food and seafood ingredients and that are mainly cooked in the most traditional Tianjin style. In addition, the restaurant offers Sichuan style dishes, Shangdong style dishes, and Cantonese style dishes to meet the needs of its customers. Its notable dishes include "steamed hairtail", "jumping carp in fishing net", "aubergine chrysanthemum", "four treasures made of diced fish", "stir-fried shrimps", "tofu with 8 delicacies".

Steamed Hairtail
This notable dish of Baoxuan's uses fresh hairtail for steaming. When cooked, the dish produces a white silver colour with a savoury and umami taste (not fishy at all) and tender texture.

Jumping Carp in a Fishing Net
Jumping Carp on the Fishing Net is a traditional dish in which fresh carp is cooked using the combined methods of "frying" and "liu". When frying, the carp is made into a curled-up shape to look like as if it is struggling within a fishing net, hence it was given the name. This dish has a few distinctive characteristics including crispy bones, tender meat, and a strong sour and sweet taste. When served, boiling sauce would be poured all over the fish,

creating a pleasant smell and a sizzling sound, providing enjoyment of sight, sound, smell and taste.

Hong Qi Shun

Hong Qi Shun is a "time-honoured brand" specialised in Muslim dishes. The restaurant has been reviving traditional Muslim dishes as well as developing new dishes in this category since its establishment in 1935. Now, the restaurant has over 1,000 different types of dishes to offer to its customers. The most notable ones are "yellow braised beef", "roast ox tongue and oxtail", "Pa seafood and mutton", and "whole mutton soup".

Red Braised Ox Tongue and Oxtail

Red braised dishes are a specialty of Tianjin's Muslim food. The "red braised Ox tongue and Oxtail" is a representative dish of the red braised variety and a popular main course of a Muslim themed banquet. The dish has a glossy red brown colour with a salty and umami flavour. It tastes savoury, sweet, pure, and rich with an aroma. The main ingredients of this dish are the middle part of an Oxtail and Ox tongue. It takes a few key steps to cook this dish:

1. Cut Oxtail into sessions, slice Ox tongue, and soak them in cold water.
2. Boil the Oxtail and the Ox tongue in boiling water in a wok until they are fully cooked.
3. Stir fry spring onions, ginger, garlic, and anise with a small amount of oil in the wok until aromatic, add in sweet flour paste (Tian Mian Jiang), plunge the Oxtail and the Ox tongue in and continue with stir frying, seasoning with broth and caramel.
4. Simmer on low heat for a while, add in salt and sugar. Continue to simmer on low heat until broth is thickened.
5. Give a final touch to the cooking by adding corn starch solution, Sichuan pepper oil, sesame oil to the dish before serving.

Deng Ying Lou Restaurant

Deng Ying Lou Restaurant is another "time-honoured brand" in Tianjin with more than 100 years of history. The restaurant is specialised in Tianjin style main course dishes, Shandong style main course dishes, folkish dishes, staple food dishes, and street food dishes. It is a designated restaurant for international tourists.

Deng Ying Lou offers to the public many notable dishes with Tianjin local flavours including "stir-fried fish slices in zao sauce", "pan-grilled prawns", "sour and spicy carp", "Shangdong style braised pig intestine in brown sauce", and "pork tripe with garlic stew".

Shandong style braised pig intestine in brown sauce
The "Shandong style braised pig intestine in brown sauce" is a chef's special in Deng Ying Lou Restaurant. This dish is well-known for its exquisite skill of building the 9-layered intestine. Firstly, divide a whole set of pig's big intestine roughly into 8 or 9 sessions, 5 inches each. Secondly, lay one session into another to start the building process. Thirdly, insert the third layer into this double-layered session. Repeat this process until there are 8 or 9 layers. Now it is the time for cooking. During the cooking process, 9 seasonings would be used, and the wok would have to be tossed for 8 to 9 times, in order for flavours to be fully steeped by the 9-layered intestine. Therefore, the Chinese name of the dish is literally meaning 9-time tossed intestine.

This multi-flavour pork intestine is elaborately made by a series of cooking techniques including boiling, deep-frying, and stewing. It is ruddy, tender with a unique blending of five tastes of sour, sweet, bitter, spicy, and salty. The hint of bitterness is provided by caramel, the spiciness is created by pepper, and the sourness of vinegar is great for balancing the original flavour of pig intestine. This dish has a pleasant aromatically sweet taste and a smooth and tender texture. It is rich and sticky but not greasy.

Zheng Yang Chun Tianjin Roast Duck Restaurant
Zheng Yang Chun is a well-known and popular restaurant brand around the country. Chairman Mao visited this restaurant in 1958 and congratulated its chefs for their superb culinary techniques and skills.

Tianjin Roast Duck
The famous "Gold-medal Roast Duck" is cooked with the traditional techniques. Roasted with fruitwood, it is one of the best to offer by Tianjin chefs. Pedigree Peking duck would be used for this dish. Fed with balanced and healthy food, the adult Peking duck would have sleek white feathers, short wings, a long back, strong and short legs, plump breasts, a stout body, tender muscles, and an even layer of fat. The gross weight of one Peking duck would roughly be 3.25kg. The preparation of the duck for roasting involves 13 meticulous procedures such as gutting and air-drying. When roasted, the uniquely flavoured duck is burgundy in colour with crispy skin and tender meat, fatty but not greasy. The slight fruity smell of the fruitwood would have covered every inch of duck.
Obviously there are many restaurants in Tianjin offering traditional Tianjin style dishes, for example, Jin Cai Dian Cang (the Collection of Typical Tianjin Dishes) Restaurant, Tianjin Dish Restaurant, and Hua Xia Restaurant.

Part One

Among many popular dishes, there are the "spring flatbread with fried shrimps", "pan-roasted prawns", "yi pin tofu" (the top-class tofu), "Pa vegetables" and "explosive stir fried lamb organs".

Spring Pancake with Fried Shrimps

This is the favourite dish of Tianjin people in spring and summer. Freshwater shrimps have a high content of protein and low content of fat. With its rich nutrients and tender texture, it could be easily digested. Its high content of magnesium can also protect the cardiovascular system. Fried freshwater shrimps would be wrapped with thin spring pancakes. The outer surface would be crispy while the fillings would be soft, creating a nice texture when eaten.

Pa Vegetables

Pa Vegetables is one of the most famous dishes at a Tianjin style vegetarian banquet. The main ingredients are vegetables, plants, grains, and mushrooms. In compliance with the strictest Buddhist food laws, there are no herbs or spices with a stimulating effect would be used to seasoning this dish. Hence, it is also called "Dish for Arhats".

The cooking method of Pa involves a unique wok skill, called Da fan shao (the big flip). It is a highly acclaimed wok tossing skill of Tianjin Cuisine and requires years of practice.

The selection of ingredients for this dish entails careful consideration to create a harmonious combination of colours. The main ingredients are shiitake mushrooms, carrot, bamboo shoot, pak choi, baby corn, lotus seed, gluten balls, oyster mushrooms, monkey head mushrooms etc. Usually, it takes at least 8 ingredients to make this dish. The ingredients would be cut into batons, pieces, and slices and then to be arranged neatly in a pattern in a wok for cooking. Simmering on low heat will help enhancing the flavour. Da fan shao (the big flip) skill will be used in the process. By flipping the ingredients in the wok skilfully, not only the pattern will remain undisturbed, but also both sides of the ingredients will be cooked evenly.

Explosive Stir-fried Lamb Heart, Liver, and Kidney

It is one of the four most famous folk dishes of Tianjin Cuisine. It might be a simple dish but requires sophisticated skills in heat control and the use of oil. Only experienced chefs can produce this dish to a level of success. At Jin Cai Dian Cang, chefs use the traditional technique to stir-fry lamb heart, liver, and kidney slices rapidly on high heat without adding in any minor ingredients, thicken the dish with corn starch solution before plating. This short yet explosive stir-frying process on hight heat illustrating years of experience. This dish has a salty and umami taste, a glossy red appearance, and a strong aroma of garlic, a perfect example of the traditional taste of Tianjin Cuisine.

4.2 Famous Tianjin Street Food Food

There are numerous varieties of street food in Tianjin. The folkish delights category alone would have more than 1,000 types of food. The three nationally well-known Tianjin street food are "Goubuli Baozi" (steamed buns), "Gui Fa Xiang Mahua" (fried dough twists) and "Er Duo Yan Zha Gao" (Ear-hole fried sticky rice cake).

Table 8 The Varieties of Tianjin Street Food

The Varieties of Tianjin Street Food	Examples
Steamed food	Zheng Bing, steamed flatbreads made of plain dough Yang Cun Gao Gan, Yang Village rice cake Lama Gao, rice cake
Sticky rice food	Ma Tuan, sesame balls Liang Guo, sticky rice cake with dried fruits Qie Gao, literally means cut cake, a type of sticky rice cake with red bean paste Tang Yuan, sticky rice balls
Pan-baked or pan-fried food	Shāobing, baked flatbreads made of leavened dough Tie Bobo, cornmeal flatbread made of leavened dough Jian Bing Guo Zi, the Tianjin style pancake
Deep fried food	You Tiao, deep fried dough sticks Guobi, crispy flatbread Juan Quan, Tianjin style spring rolls Zha Gao, fried rice cakes Mahua, fried dough twists
Soup-based and gravy-based foods	Wonton soup Lao Tofu, silken tofu with gravy Dou Jiang, soya milk Gabacai, sliced jiangbing (pancake) served with gravy Mian Cha, a type of porridge made of millet flour or sticky millet flour seasoned with sesame paste, sesame oil, sesame seed, and salt Xiao Dou Zhou, a type of congee made of various beans and grains
Food with fillings	Baozi, steam buns with filling Jiaozi, dumplings Shao Mai (also Romanised as Siu Mai for a Western audience), a form of dumplings Guo Tie, fried dumplings Hezi, a round and flat shaped dumplings Xian Bing, similar to closed Pizza
Folkish delights	Lizi Geng, chestnut paste Qiu Li Bai Tang, sugar-pickled Autumn pears Su Candou, baked broad beans Qiangmi Nuo, lotus root filled with sugar-marinated glutinous rice, Liangfen, jelly noodles made of sweet peas starch Shui Bao Du, water-launched lamb or beef tripe slices Yang Tang, mutton soup.

Many of the street food varieties have won national awards. Among which, 43 were awarded the title of "the most famous street food of China".

Part One

Tianjin Baozi

Tianjin baozi (steamed buns with filling) is an affordable and popular street food range that enjoys a high reputation regionally, nationally, and internationally. The most representative bun of all types is the "Goubuli baozi". "Goubuli baozi" is one of the three best Tianjin street food. The recipe was created in 1858, more than 100 years ago in the Qing Dynasty.

According to the tale, a baby boy was born into a peasant's family in Zhu Village, Wuqing County, Shuntian City in Hebei during the reign of Emperor Dao Guang.

The father, who was then in his forties, gave his son a low baby name Gouzi (dog) to wish his child a long and peaceful life. It was believed that a low baby name can ward off evil spirits.

Years later, the poor family condition forced the father to send Gouzi to learn Baozi making skills as an apprentice in Tianjin. Three years later, when he finished his apprenticeship, he decided to start his own business with his own recipe. Gouzi set up a simple tent near his master's house to make and sell baozi. His Baozi was so tasty that people would queue for a long time to buy it. Gouzi was too busy to talk to his customers. After a while, his customers started to call him Goubuli, which literally means "dog ignores me".

Headquartered in Tianjin, the Goubuli company has opened more than 80 branches across the nation. There are also authorised Goubuli franchises in the United States, Japan, South Korea, and Singapore. Goubuli Baozi used to have only four varieties of fillings, and then the number of filling varieties had grown into over 20. Nowadays, there are 10 ranges of Baozi products with more than 100 different kinds of fillings. Baozi themed banquets are offered at different price levels for different customer groups.

Besides the "Goubuli baozi", "Shitou Menkan su bao" (Stone Threshold Vegetarian Buns) is also very popular in Tianjin. It is another Tianjin street food with more than 100 years of history. The shop that produced the vegetarian buns was originally called "Zhen Su Yuan" (literally means authentic vegetarian shop). The story has it that Zhen Su Yuan was flooded every time when there was a heavy rain due to the poor rainwater drainage outside the shop. Hence, a stone threshold was built to prevent the rainwater from coming into the shop. As time passed by, the shop's original name was forgotten, and people would always refer it to as the "Shitou Menkan" (Stone Threshold).

Chinese Food Culture

Tianjin Mahua (Fried Dough Twists)

The most famous brand of "Mahua" (fired dough twists) in Tianjin is "Gui Fa Xiang", one of the three best street food of Tianjin. 70 years ago, there were not many varieties of dough twists in Tianjin. As a result of the business competition against his brother, Fan Guilin improved the recipe to have created two new flavours of dough twists in 1937. One had added filling into the dough and the other was made of half-leavened dough. Recipes would be adjusted according to the change of season and climate so that the quality of the dough twists would remain consistent.

Gui Fa Xiang fried dough twists are crisp, short, aromatic, and sweet. It has a long shelf life. in other words, it can be stored for a long time without becoming soft.

Tianjin Zha Gao (Fried Sticky Rice Cake)

"Zha gao" literally means fried cake. It is a traditional and easily accessible street food in Tianjin. "Ear-hole fried cake", along with "Goubuli baozi" and "Gui Fa Xiang Mahua" are known as "The Three Treats of Tianjin".

"Ear-hole fried cake", originated in the Qing Dynasty under the regime of Emperor Guang Xu (1875 to 1908), was founded by Liu Wanchun as a family business. In the early days, he used to walk along the streets selling fried cakes from a wheelbarrow. Later, he partnered with his nephew Zhang Kuiyuan to open the Liu's Fried Cake Shop. The Chinese word "cake" is a homophone for "height" which has an auspicious meaning of higher achievements. Hence, the fried cake was popular among the rich and the poor especially on occasions of wedding, birthday celebration, and funeral. Liu's cake became so popular to the extent that customers had to order in advance. Liu's shop was next to the Ear-hole Alley. It was a long and narrow alley, slightly more than 1 meter in width. People simply referred to Liu's fried cake as the "Ear-hole fried cake" rather than calling it "the fried cake from Liu's Fried Cake Shop".

The traditional method of making Ear-hole fried cake involves grinding pre-soaked proso millet and glutinous rice into thick porridge with millstones. The liquidised grains would be placed in a cloth bag to be rinsed under running water and then set aside to leaven. Once it is leavened, adding food grade sodium bicarbonate to the dough, it would be ready to use. To prepare the red bean filling, start with making paste using local red beans. The red bean paste would then be stir-fried with good quality dark brown sugar in a wok. When paste cools down, wrap a small amount of red bean paste with a wrapper made of the grain dough. Heat oil up on medium heat, place cakes in, flip them from now and then. The finished cake would have a golden-yellow colour and a rough surface. It would be crisp and short on the outside and tender and sticky on the inside. The red bean paste filling would taste smooth and sweet with a unique flavour.

Part One

Jianbing Guozi (Tianjin Crêpes)

Originated in Tianjin, "Jianbing" is possibly China's most popular street food. The thin batter is made of mung bean flour and seasonings. A delicious Jianbing would be cooked following the steps below:

1. Deftly spread the batter into a crêpe.
2. Crack an egg or two onto the uncooked surface and spread evenly, sprinkle chopped spring onions, coriander, and sesame seeds.
3. Flip the crêpe, place a youtiao (fried dough stick) or guobi (crispy flatbread) on top, spread tian mian jiang (sweet flour paste), furu (fermented bean curds) or la jiang (chilli sauce), fold it up and it is ready to serve.

Jiangbing has an aromatic and salty flavour and a tender and crisp texture.

Nowadays, mung bean flour is no longer the only choice of the main ingredient to make Jiangbing. Other options include soya bean flour, black bean flour, or mixed grain flour etc. However, Tianjin people still prefer the traditional recipe, using mung bean as the main ingredient, add youtiao or guobi, spring onions, sesame seeds and other seasonings.

Tianjin Jiaozi (Dumplings)

Dumplings are a year-round favourite food to people in Tianjin. This tradition can be explained by the local food customs and climate of Tianjin.

From the food customs perspective, making and eating dumplings is a common means to celebrate festivals, holidays, family gatherings, and birthday. This custom is best represented by the tradition of eating dumplings on Chinese New Year's Eve. When treating families and friends, Tianjin people serve dumplings. On the wedding day, parents would send their daughter off with dumplings. The day prior to a birthday, dumplings would be served as a prologue of the birthday celebration.

Tianjin people also have the tradition to eat dumplings at the beginning of the "San Fu" period which is the hottest 40 days of a year. "San" means three and "Fu" means to hide for the summer heat. The 1st Fu and the third Fu each lasts for 10 days. The middle Fu is a period of 20 days. San Fu occurs between the 12th solar term "lesser heat" and 14th solar term "end of heat" in Chinese lunar Calendar.

Bear the best wishes for life, all these traditions are aiming to create some fun in life as well.

Tianjin people are meticulous about the filling of dumplings. There are three varieties of fillings including vegetarian, meat, or mixed to cover a range of combinations of ingredients and flavours. Meat fillings are made of a single type of meat ingredient. Vegetarian fillings would be divided into vegan and vegetarian. For vegetarian dumpling fillings, small sized dried shrimps, eggs, spring onions, ginger, and garlic can be use whilst they are prohibited ingredients for vegan fillings. As its name suggested, the mixed fillings are made of a combination of meat and vegetables.

There are also other forms of dumplings such as "shao mai" (also known as siu mai) in the shape of a pomegranate, or Hezi, which use two wrappers to create a round and flat shape of dumplings.

Longxu Mian (Dragon Beard Noodles)

Dragon Beard Noodles are the traditional varieties of Northern style dough food. It is a spin-off product of hand-pulled noodles that originated in Shandon province with a history of more than 300 years. It was said that in the

Ming Dynasty, there was a chef from the royal kitchen who made hand-pulled noodles on the first day of Spring that was as thin as hair, just like the beard of a dragon. The Emperor could not stop praising it while eating it. Since then, this fried thin noodle dish has become a trendy dough food. Since it was made with great effort, and the noodles look like intersecting dragon tentacles, hence it was named after it.

Tianjin Xishi (Soup-based and gravy-based foods)
Tianjin Xishi is soup-based and gravy-based foods that are filling. They are typical breakfast choices. The most common dishes in this category are wonton soup, guobachai, and silky tofu with gravy.

- Tianjin wonton soup

The filling of Tianjin wonton is made of minced pork with seasonings. Comparing to the dumpling wrapper, the wonton wrapper is thinner and in a square shape. Wonton would be cooked in boiling water until fully cooked. To serve, place some xiapi (dried small shrimps) and dongcai (preserved mustard greens) or roasted seaweed in a soup bowl, add in broth, wonton dumplings, and chopped coriander. Tianjin wonton soup has a unique flavour of umami and aromatic flavour. For vegan wonton, shiitake mushrooms, rehydrated dried bamboo shoot slices, beansprouts, and chopped coriander would make delicious savoury filling. In winter, clay-pot wonton soup is popular. It is umami, aromatic, and hot in temperature. A perfect hearty dish for winter!

- Silken tofu with gravy

To make Silken tofu, grind pre-soaked soya beans with millstones to obtain soya milk. Bring soya milk to the boil in a pot, add coagulant (curdling agent) such as Yan Lu (nigari salt solution) or gypsum, continue to simmer soya milk until it becomes solid. Silken tofu is the softest variety in the tofu family.

"Silken tofu with gravy" is a DIY type of dish to suit individual taste. There would be a range of options of gravy and toppings to be served with the freshly made Silken tofu. Customers can create their own flavours by combining the grave of choice with different toppings. The options of gravy include standard style gravy, shrimp gravy, mutton and mushroom gravy, muxu pork gravy, and halal grave.

- Gabacai

Gabacai, also known as guobacai, is a unique variety of Tianjin gravy-based food. Da Fu Lai is the most well-known Gabacai brand. The main ingredient of this dish is a type of Jianbing (pancake) that made of millet and mung bean flour. Pancakes would be sliced into willow leave shape and air dried. These pancake slices would be served in vegetarian gravy and topped with sesame paste, fermented bean curd solution, chilli oil, chilli paste, five-spice dried tofu slices, and chopped coriander. Gabacai has a refreshing amora, a crunchy, tender, and slightly chewy texture, glossy and runny gravy.

Part One

Tianjin Sticky Rice Food

Traditionally, sticky rice food was a type of delicacy designed for summer months. As its name suggests, the main ingredient of this type of food is sticky rice, also known as glutinous rice.

Tianjin sticky rice food has a history of 300 years. There are a wide variety of choices. Sticky rice would appear a little translucent once cooked, a perfect ingredient for creating cakes, balls, and many desserts with a tender, silky, sticky, chewy, smooth, and moist texture.

It is worth mentioning that the Zhilan Zhai rice cake is a specialty of Tianjin. The cake is very presentable, tender, chewy, and moist. Although it contains sticky rice, it never gets stuck on the teeth. It can be served cold or hot. A truly irresistible Tianjin delicacy.

Part Two: Ingredients & Equipment

Chapter 7 Ingredients & Equipment

Section 1 Commonly used ingredients in Chinese Cooking

Cooking Wine
Based on the origin of the produce, there are Shaoxing wine and yellow wine with the latter being widely used for cooking.

Chinese Soy Sauces
There are many varieties of Chinese soya sauces depending on the variation of consistency, aging and strength of the flavour. The most widely used varieties are *fresh soy sauce* (light soy sauce) and *aged soy sauce* (also known as dark soy sauce or old soy sauce) both originated in Guangdong and Guangxi provinces. Before the fresh soy sauce and aged soy sauce became popular, people in the north of China used to cook with **northern**

style soy sauce. Additives with sweet or umami tastes are sometimes added to a finished brewed soy sauce to modify its taste and texture. This type of soy sauce variety is labelled as **seasoned soy sauce**, for instance, Mushroom dark soy sauce, seasoned soy sauce for seafood.

Northern Style Soy sauce originated in northern China. It has become less popular. Hence, there are not many manufacturers producing this variation as it used to be. Among the few widespread trademarks, "Glory" (in glass bottle) is produced by Tianjin Limin Condiment Co. Ltd. It is a thin, medium-brown soy sauce with slightly salty and umami tastes. It has an intense fermented soy flavour and aroma.

Fresh soy sauce, more commonly known as light soy sauce is a thin, opaque, lighter brown soy sauce. It is saltier and stronger in umami taste with an intense fermented soy flavour and fragrance.

Aged soy sauce, more commonly known as dark soy sauce. A darker and slightly thicker soy sauce made from light soy sauce. It has a richer, slightly sweeter, less salty, less fermented soy flavour and less aroma. It is usually used to add colour to a dish.

In cooking, northern style soya sauce is normally used by itself while light and dark soy sauces are likely to be combined together. Seasoned soy sauces are good supplementary to the other three varieties.

Vinegar

There are many variations of vinegar in China. Sichuan cuisine is in favour of Baoning Vinegar, many northern regions prefer Shanxi Extra Matured Vinegar, Zhenjiang Fragrant Vinegar is popular in most of the southern regions. Tianjin Cuisine tends to select Duliu Aged Vinegar. In Chinese cooking, vinegar is referred to as coloured vinegar (such as aged vinegar and fragrant vinegar) and clear vinegar (rice vinegar).

Sugar

The most commonly used type of sugar for cooking in China is white sugar with a soft texture, which isn't readily available in the UK; White castor sugar is the preferred alternative.

Pixian Board Bean Paste

It originated in the county of Pi in Chengdu, Sichuan, China, made from fermented Sichuan Erjingtiao red chilies, board beans and spices.

Sichuan Style Pickled Red Chilies

It is made from fermented Sichuan Erjingtiao red chilies and spices.

Pepper

There is a tradition of using white pepper in Chinese cooking. As an influence the western and other cuisines, black pepper is used in some dishes. It is widely believed that white pepper would help to remove meat odour.

Chili Peppers

There are many varieties of chili peppers in China. Famous ones include "Beauty chilies" from Hunan Province, "Millet chilies" (a variation of bird's eye chilies) from Yunnan Province, "Erjingtiao chilies" from Sichuan Province, "Qin chilies" from Shanxi Province, and "Tianying chillies" from Tianjin.

Sichuan Peppercorn

This is known as Chinese prickly ash of which there are two variations:

Ingredients & Equipment

Flower Peppercorn

It is red brown in colour, can be found in most of the provinces in China except for Guangdong, Hainan, and Taiwan. It has a unique aroma and taste and creates a special sensation (a kind of numbness) in the mouth.

Ma Peppercorn

This green Sichuan peppercorn is only produced in Sichuan and Guizhou. Comparing with flower pepper, it creates the same sensation in the mouth but much stronger.

Sesame Oil
It is also called fragrant oil. It is made from sesame seeds and has an intense aroma. It is used to add flavour and aroma.

Sichuan Peppercorn Oil
It is produced by frying Sichuan peppercorns in vegetable oil. In most cases, ma peppercorns variation would be used. The oil is an essential in cooking dishes with a combined flavour of ma and xiang (mouth numbing and fragrant).

Oyster Sauce
Originated in Cantonese cuisine, it has become popular with all styles of Chinese cuisine. It is made from fresh oysters to give a salty, sweet, fragrant, and a slightly sour flavours. Popular brands include HADAY, LEE KUM KEE.

Tian Mian Jiang
The literary translation from Chinese to English is Sweet Flour Paste, or Fermented Flour Paste. However, it might be known outside of China as sweet bean paste due to mistranslation even though the main ingredient of it is flour, not beans. It creates salty and sweet flavours and fermented flour aroma. It is the sauce used in eating Peking Roast Duck.

Chinese Star Anise
Chinese star anise is widely used in Chinese cuisine in which star anise is used to enhance the flavour of meat. It is an ingredient of the traditional five-spice powder.

Fermented Tofu
Fermented tofu is known as Furu in China. It is also referred to as fermented bean curd, or soy cheese in English speaking countries. Depending on the origin of produce, Furu would look and taste differently. Southern style,

109

for example, has an ivory colour with a stronger fermented soya taste. Sichuan style uses chilli oil to enhance the flavour. Beijing style is red in colour with a salty and aromatic flavour. Regardless of the distinctive features, fermented bean curd in general tastes salty, umami with some mild sweetness in it. The texture and taste of Furu resembles a firm, smooth paste similar to creamy blue cheese. Its culinary use is to add colour and flavour to dishes.

Shrimp Sauce
It is made from fresh shrimp. It is salty and umami.

Dry-cured Chinese Hams.
Dry-cured hams are considered regional delicacies. Jinhua, Xuanwei, and Rugao are three areas that produce the best hams in China.

Bamboo Shoots
As the name implies, bamboo shoots are the edible shoots of the bamboo plant, which is native to Asia. They are cut from the plant once they appear above the ground to preserve their tenderness and because if they are left to grow exposed, they will turn a green colour.
Fresh bamboo shoots are available at Asian or Chinese markets, or you can find canned bamboo shoots at most local grocery stores. Fresh shoots need to be boiled until tender, then baked and cut into pieces. Canned bamboo shoots only need to be heated since they are pre-cooked.

Bitter Melon
Bitter melon is known for its unusual appearance and taste. This Chinese gourd resembles a cucumber with a dark green, bumpy, pockmarked skin. As the name implies, it has a rather bitter taste; however, this can be lessened by blanching or disgorging the melon with salt.
Bitter melon is a popular ingredient in stir-fries, such as Chinese pork with bitter melon, where it is frequently paired with other strong flavours, such as Chinese salted black beans.

Bok Choy
China's most popular vegetable, Bok choy, has a light, sweet flavour and crisp texture. It is a type of cabbage, but instead of a tightly packed head, the leaves are in a cluster, giving the vegetable a shape similar to celery.
Bok choy (also called pak choi) is used to enhance everything from potstickers to to stir-fries. Nutritionally, like most leafy green vegetables, bok choy is a good source of iron; it is also high in Vitamin A, Vitamin C, and calcium. It should be stored in the refrigerator in an air-tight container for 3 to 4 days.

Gan Lan
Gan lan has a slightly bittersweet, earthy flavour that pairs nicely with strongly flavoured ingredients such as oyster sauce. It is best to sauté, steam, or stir-fry this vegetable.

Ingredients & Equipment

Chinese Celery

Regular celery can seem a little boring once you've tried Chinese celery. Originating from a wild celery native to Asia, Chinese celery has a significant flavour that adds extra taste to soups and stir-fries. It is not recommended in raw salads though, as its strong flavour may overpower other ingredients; cooking this vegetable mellows the flavour.

Chinese celery also has a more attractive appearance than regular celery, with thinner stalks and a colour ranging from dark green to white. The stalks are thinner than the celery we are used to but do need to be crushed slightly before adding to a recipe. This will improve their texture and flavour.

Dried Fragrant Mushrooms

Dried fragrant mushrooms, or dried shiitake mushrooms, are a major source of umami taste which makes it a signature ingredient of Chinese cuisine. Although available fresh, fragrant mushrooms are most often found dried to use. They need to be rehydrated by soaking in the water to soften prior to cooking and the soaking liquid is often used in replacing of water or chicken broth in cooking.

Black-ear Mushrooms and Silver-ear Mushrooms

These mushrooms are dried to use in both cooking and natural medicine; practitioners of traditional Chinese medicine believe dried mushrooms have a cleansing property and can help lower blood pressure.

Chinese Aubergine

Chinese aubergine resembles more of a purple courgette, with its long thin shape and purple colour that may be streaked with white. Aubergine is native to Asia, and Chinese aubergine is one of hundreds of varieties found there (another type of Southeast Asian aubergine, Thai aubergine, is small, round and green or white in colour).

Because it is smaller and thinner than the usual oblong-shaped aubergine you find in the produce section of local supermarkets, Chinese aubergine is not normally salted and disgorged before cooking. Popular Chinese dishes made with aubergine include Sichuan aubergine braised in garlic sauce and aubergine in garlic sauce.

Silk Gourd

Silk gourd, also called Chinese okra and angled luffa, is a long thin gourd with textured skin. Only immature silk gourds are eaten, as older silk gourds have a bitter taste. Silk gourd can be stuffed with pork and steamed; however, it is more commonly stir-fried or deep-fried.

Feel free to substitute silk gourd in recipes calling for cooked zucchini or okra, like in a stir-fried okra recipe, and to use okra as a substitute if silk gourd is unavailable.

111

Chinese White Radish

Chinese white radish, also known as mooli radish, daikon, resembles a large white carrot. It has a much stronger flavour than small, round red radishes. While it is a popular salad ingredient in Japan, in China it is more commonly used in cooking, both in stir-frying and slow-cooked dishes; it is also pickled and added to soups.

Nutritionally speaking, Chinese white radish is a good source of Vitamin C and is very low in calories.

Hairy Gourd

It goes by many names such as fuzzy melon, hairy cucumber, hairy squash. In north of China, it is known as wo gua. If Chinese aubergine looks like a large purple courgette, hairy gourd looks like a courgette covered with fine hair. However, while courgette is a type of squash, hairy gourd is a gourd, related to winter melon. In Chinese cooking, hairy gourd is used in a number of dishes such as soups and stir-fries. It can also be filled and steamed. When choosing hairy gourd, look for ones that are small and firm. Peel off the skin or scrub well to remove the "hair" before using.

Winter Melon

Winter melon, or dong gua, belongs to the gourd family. It is a versatile vegetable, often used in stir fired dishes, braised dishes, soup dishes, desserts, and baozi or dumpling fillings. It is also a popular choice for vegetable carving. The famous "Winter melon Zhong" (winter melon container) is made of winter melon.

Chinese Garlic Chives

Although both are members of the onion family, Chinese garlic chives, or jiu cai, are more attractive and more flavourful than the more recognisable chives used as herbs in western cookery. Chinese garlic chives are larger in size, leafy, garlicky in flavour, and juicy and crisp in texture. It is considered as a type of vegetable rather than garnish to Chinese people. Chinese garlic chives are a regular main ingredient in fillings for dumplings, spring rolls, and steamed buns. It is also used for adding flavour to stir fried dishes.

Chinese Yellow Chives

Thought to be more of a delicacy than jiu cai, Chinese yellow chives, or jiu huang, are chives that have been grown under cover without exposure to direct sunlight (hence the yellow pigment). Yellow chives are a tender version of garlic chives. The taste may be essentially the same, but the texture and flavour are far more delicate.

Flowering Chives

Flowering chives, or flowering Chinese chives, is named for their hollow, light green stems and yellow buds, these stalks are light and crisp. Flowering chives is less garlicky than either the garlic chives or yellow chives.

Ingredients & Equipment

Chinese Leeks

Leeks commonly found in supermarkets have a mild sweet flavour. Chinese leeks, on the other hand, are smaller and thinner, resembling a thick spring onion. In China, it is known as da cong, literally means large spring onion. Their more juicy and pungent flavour makes Chinese leeks a staple ingredient in northern Chinese cooking. Chinese leeks generally are not easily available in the UK, so regular leeks can be used as a substitute.

Long Beans

Believed to have originated in China, long beans come by their name honestly as they can grow up to three feet long. Other common names for long beans include yard long beans, snake beans, and Chinese peas. Long beans are a member of the same family as black-eyed peas, also called cowpeas.

In cooking, long beans are the beans traditionally used to make Chinese green beans, a popular dish at Chinese buffets. They are also popular in Southern China, and Cantonese cooks frequently pair long beans with salted black beans or fermented bean curd. Outside of China, long beans are used in Malaysian and Thai cooking.

Lotus Root

A relative of the water lily, the lotus is an aquatic plant that grows in marshes and shallow ponds. The tuberous root of the plant is found in the mud below the surface. The exterior of the root is not particularly attractive, resembling a large, buff- coloured link of sausages, with each link about 8 inches long. However, channels running through the root give cut slices a delicate, lacy appearance, and lend itself to a stuffed recipe such as lotus root stuffed with sweet sticky rice.

Lotus root adds a crisp texture and sweet flavour to Chinese stir-fries, soups, and salads, where they are often added raw. Red-cooked pork belly with lotus root and kinpira renkon (braised lotus root) are two recipes where the lotus root is cooked. Deep-fried lotus root is a popular garnish. You'll frequently find candied lotus root in Chinese New Year Trays of Togetherness (also called Harmony Trays), as a symbol of abundance.

Mung Bean Sprouts

Chinese people have been sprouting mung beans for approximately 3,000 years. The Chinese name for the silver-coloured sprouts with the yellowish ends is dou ya cai. Their crisp texture and sweet flavour are used in stir-fries, salads, noodle soups, and fillings.

Nutritionally, mung bean sprouts are low in calories, fat, and carbohydrates, and high in Vitamin C, and a good source of protein. Stir-frying the sprouts helps reduce the chance of food-borne illness; however, to preserve their crunchy texture, mung bean sprouts should not be stir-fried for too long.

Chinese Leaves

There are hundreds of varieties of cabbage used in Asian cooking. The cabbage, commonly known as Chinese leaves is the large cabbage with the pale green leaves.

Chinese leaves have a mild, sweet flavour that pairs nicely with more strong-flavoured foods. It adds texture and flavour to stir- fried noodles,

113

dumplings, soups, and hot pot broth. It can be steamed, stir-fried, and sauteed.

Snow Pea Shoots
Snow pea shoots, or dou miao, are the delicate tips of the vines and the top set of leaves of the snow pea plant. Considered to be a delicacy in Chinese cooking, snow pea shoots can be served raw in salads, simply sauteed, quickly cooked in stir-fries, or blanched and used in soups.

Chinese Isinglass
Chinese Isinglass, or Chinese gelatine, or agar-agar, is a powdered gelatine substitute made from algae that does not require refrigeration. It is a vegan alternative to gelatine, which is an animal product.

Chinese Five-Spice Powder
Use equal amounts of cinnamon, star anise, cloves, fennel, and Sichuan peppercorn. If Szechuan peppercorns aren't available, use freshly ground black peppercorns. If you need five-spice powder for slow cooking meat, then you can just use one or two-star anises, cinnamon sticks, and cloves as a substitute for the five-spice powder.

Coconut Milk
Coconut milk is a little thicker and creamier than regular milk, and some recipes will benefit from the flavour it adds.

Seaweed
There are different types of seaweed. In Chinese cuisine, seaweed is used mainly for salads, braised dishes, or soups.

Rock Sugar– Bing Tang
Rock sugar, or crystal sugar, or bing tang. It is a key ingredient in many dishes, including "braised pork belly", "red cooked pork", "sweet and sour ribs", and "Chinese braised oxtails". It adds the sweetness and the perfect gloss on meat dishes.
Nowadays, rock sugar is sometimes replaced with caster sugar.
Rock sugar also comes in brown sugar pieces and is call bing pian tang or literally rock sugar slices.

Chinese Liquorice Root
Chinese Liquorice root or gan cao is a very common ingredient used in Chinese herbal medicine, but it is also used to flavour broths and braised dishes.

Sesame Seeds - White & Black
Sesame seeds or zhi ma are an essential pantry item for Chinese cooking. Sometimes used for sweet treats and sometimes for accent or garnish, they add a great nutty flavour and crunch. There are various colours of sesame seeds available and both white and black seeds are used in Chinese cooking.
It is best to store sesame seeds in a tightly covered container in the refrigerator after opening, but if kept in a dry and cool area, it also keeps well in the pantry. Sesame seeds are quite heat tolerant, it's actually not easy to burn sesame seeds.

Ingredients & Equipment

Dried Tangerine & Mandarin Orange Peels
Dried tangerine and citrus peels, commonly referred to as guopi or "fruit skin" in Chinese, are a widely used spice in Chinese cooking. It is used in dishes like Tangerine beef and Orange chicken, but also in soups, stews and braised dishes like Braised Duck. You can try drying your own or buying it from your local Chinese grocery store.

Turmeric Powder
Turmeric powder or huang jiang fen is an essential ingredient in all curry powders and provides most of the yellow golden colour and is a healthy herb. Adding extra turmeric powder enhances the colour of a good curry dish and can also be used in place of food colouring when a nice yellow colour is desired.

Cumin
Cumin, or zi ran, is an aromatic seed used to flavour sausages, lamb, squid, and a variety of Chinese dishes.

Fennel Seed
Fennel seed or hui xiang zi is one of the spices used to make 5 spice powder. Whole fennel seed are used in Lanzhou beef noodle soup and many stewing dishes.

Chinese Cinnamon
Also called cassia cinnamon, this is unlike regular cinnamon in colour as well as thickness. The more common name in Chinese is gui pi but it is also known as rou gui.
Chinese cassia cinnamon is thicker, greyer, and more closely resembles tree bark. This is also one of the main ingredients in five-spice powder.

Cloves
Cloves or ding xiang have very strong pungent, and sweet aroma and is one of the five spices that make up powder. It can be used whole or ground up and is widely used both Asian cooking. Whole cloves can be ground as needed or just let whole cloves infuse dishes like Homemade Chili Oil.

Black Cardamon
Cardamom or hei doukou is an ingredient that you probably think of when cooking Indian dishes, but they also have their place in Chinese and Southeast Asian cooking. Green cardamom is seldom used, but black cardamom adds an extra dimension of flavour to stews and noodle soups.

Amomum White Cardamon
Amomum white cardamom, sometimes called Thai cardamom, is used throughout China and Southeast Asia for its floral aromatic flavours. It is sometimes ground and used in curry powder but often used in braising liquids

115

Part Two

and aromatic soups. The white cardamom has a hint of menthol flavour and does wonders for adding a unique flavour to your dishes.

Dried Sand Ginger

Dried Sand Ginger (sha jiang, or shannian) is different from your regular variety of Chinese ginger and is part of the galangal family and sometimes referred to as aromatic ginger. A side-by-side comparison of sand ginger powder or sha jiang fen with ginger powder, jiang fen will find that they taste and smell quite different as they are different varieties of ginger.

Section 2 Large Kitchen Equipment

2.1 Deep Fat Fryers

Health and Safety

The oil should be kept at the recommended level.
The thermostat should be checked to ensure it is operating properly. Furthermore the deep fat fryer should never be left unattended, or near water or any type of material in case of fire.

Why Are They Used in The Industry?

Producing food in restaurants and takeaways has to be done in a quick and efficient way to keep on top of orders made. This is where the deep fat fryer comes into good use as it speeds the cooking process up many dishes.

It's believed by experts that the technique of frying foods originated in the Middle East. The reason fat or oils are used is because they can reach such a high heat. Unlike boiling water, fat or oil becomes so hot that they can sear the outside of food. This keeps the moisture and flavour in the food, but also leaves the outside crisp.

2.2 Atmosphere Steamers

Health and Safety

Store in an appropriate place in the kitchen on level ground. Allow it to cool before cleaning.

Why Are They Used in The Industry?

Atmospheric steamers offer the perfect solution for volume production of healthy steamed food. Simple to use with a generous capacity, these versatile steamers are ideal for cooking a host of other popular food products. The gentle steaming process retains important nutrients & reduced shrinkage saves you money.

2.3 Bain Marie

Health and Safety

You should Check the thermostat in a commercial called Bain Marie. The commercial Bain Marie should be stored at an easy to reach level, in an appropriate place in the kitchen. With a Bain Marie you should check that there is no material that could catch fire.

Ingredients & Equipment

Why Are They Used in The Industry?
An improvised Bain Marie is used to melt chocolate. Chocolate can be melted in a bain-marie to avoid splitting and caking onto the pot. Special dessert bain-marie's have a thermally insulated container and are used as a chocolate fondue.

Cheesecake is often baked in a bain-marie to prevent the top from cracking in the centre. Custard may be cooked in a bain-marie to keep a crust from forming on the outside of the custard before the interior is fully cooked. Classic warm sauces, such as Hollandaise and beurre blanc, requiring heat to emulsify the mixture but not enough to curdle or "split" the sauce, are often cooked using a bain- marie. Bains Marie can be used in place of chafing dishes to keep food warm for long periods of time, where stovetops or hot plates are inconvenient or too powerful.

2.4 Bratt Pans

Health and Safety
Check the thermostat regularly to ensure it is operating properly. Store at an appropriate height on a level surface.

Why Are They Used in The Industry?
Bratt Pans are a versatile piece of high production catering equipment which is perfect for large quantity production. A bratt pan has a deep rectangular cooking pot and the heat source to the cooking pan can be either gas or electric. A bratt pan will have a tilting mechanism for emptying the contents out and to aid cleaning. The tilt feature can be electrically operated or a manual hand driven mechanism. A bratt pan is perfect for braising, sealing, shallow frying and general cooking applications.

2.5 Griddle

Health and Safety
On the commercial griddle you must check the thermostat. Ensure that it is stored in an appropriate place in the kitchen and at an easy to reach level.

Why Are They Used in The Industry?
The technique of griddling foods may be used to cook with dry or moist heat, and with or without oil. Griddled foods include pancakes, oatcakes, crepes, grilled cheese, unleavened bread (roti or chapatti).

Part Two

2.6 Blender

Health and Safety

You should make sure there are no trailing wires. When washing a blender you should ensure the attachable blades are in sight on the side until they are washed. Make sure the guard is always on correctly.

Why Are They Used in The Industry?

Making smoothies, soup puree.

2.7 Mixer

Health and Safety

Make sure there are no trailing wires, and store the it on a level surface.

Why Are They Used in The Industry?

Ingredients that are in bulk, for instance making large quantities of cakes for bakeries. These are also useful for speeding up the process of making a range of different breads in Asian restaurants.

2.8 Processor

Health and Safety

Ensure there are no trailing wires, and store at height level. Make sure the safety lid is always secure.

Why Are They Used in The Industry?

A food processor is used to blend and process a variety of foods, such as fruit and veg to make smoothies and soups. They can also be used to blend numerous cakes mixtures together, to speed up the process.

2.9 Conventional Ovens

Health and Safety

Make sure there are no objects nearby that can catch fire. Always keep the flames to the size of the pans. Keep fire equipment nearby.

Why Are They Used in The Industry?

Baking, boiling, simmering, frying.

2.10 Combination Stoves

Health and Safety

Be careful of steam when opening the doors. Store on level surface at height level.

Why Are They Used in The Industry?
- Oven baking & steam

Ingredients & Equipment

- Controlling core cooking temperature
- Settings that control moisture and dry heat
- A variety of foods can be cooked
- Selfcleaned

2.11 Salamander

Health and Safety
Use oven cloth and check the thermostat

Why Are They Used in The Industry?
- To grill a variety of food.
- Stored at eye level to keep an eye on the food
- For holding food, to keep it warm without burning
- Also, for glazing food

2.12 Industrial Wok Burner Chinese Stove

Health and Safety
There may be hot large flames so need to ensure there is nothing that could catch fire near the flames.

Why Are They Used in The Industry?
Allows the flames to transfer heat to the entire surface area of the wok for more efficient food production.

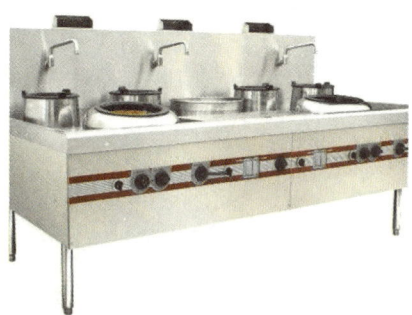

2.13 Industrial Induction Chinese Wok Cooker

Health and Safety
Should always stand at least six inches away from the device while it is being used.

Why Are They Used in The Industry?
Induction wok Cooker represents some the most advanced technology. Compared with the traditional gas or electric stove, it is safer, more convenient and reliable. With accurate temperature control and high thermal efficiency, this smoke-and-fire free appliance results in little indoor air pollution.

2.14 Rice Cookers

Health and Safety
Clean out regularly.

Why Are They Used in The Industry?
Holding rice for up to 12 hours.

119

Part Two

Section 3 Chinese Kitchen Cooking Utensils

3.1 Wok

A wok is a versatile round- bottomed cooking vessel, originating in China. Woks come in three forms.

- Wok with one handle, or chao guo

- Wok with two handles, or double-eared wok, or er guo

- Flat-bottomed wok

Deep frying and stir-frying cooking techniques require light-weighted, solid and durable wok that enables fast transmission of heat. Hence, traditional Chinese wok are usually made of wrought iron, and those with a clean, glossy, and smooth surface are the best. Iron wok should be cleaned after each use. To care for it, wok should be heated up on intensive fire once a week to remove the grime and dirt from the inside and outside. After a thorough heat up, wash the wok with mild detergent, clean it with water, dry it up on heat, and then polish it with a mix of oil and spring onions leaves to keep the inside of wok moist, glossy, and clean.

3.2 Chinese Cooking Ladle

Chinese cooking ladle, or hand ladle is a utensil used for mixing ingredients in the wok, adding seasonings, and scooping cooked food out from a wok into a plate. Although it has slightly different design for different cooking techniques, the handle is usually of 30 cm length and made of food-grade stainless steel.

Ingredients & Equipment

3.3 Chinese Scoop Strainer

Chinese scoop strainer is used for straining, skimming, and deep-frying. The diameter is about 20-26 cm. With a wooden handle, it is usually made of food-grade stainless steel. The diameter of the scoop is about 20-26 cm which makes it very convenient to lift a fried fish, pieces of meat, and chips from hot oil, or to remove noodles, wontons, and dumplings from boiling water. With a wooden handle, it is usually made of food-grade stainless steel.

3.4 Kitchen Sieves

Kitchen sieve is a utensil consisting of a wire mesh held in a frame, used for separating impurities from soup, oil, and other liquid-based seasonings. The fine mesh sieve is made of copper and is best for straining out food residues in soup. The medium mesh sieve is usually made of food-grade stainless steel and is used for filtering out coarser from finer particles or lumps in oil.

3.5 Wok Shovel

A wok shovel is designed for flipping and scooping during the cooking process. To cook different dishes, there are round-edged shovel and square-edged shovel to choose from. With a wooden handle, the mini shovel is about 35cm in length and is made of food grade stainless steel.

3.6 Oil Drum

Oil drum is a type of container to hold cooking oil. With a capacity of 5kg, an oil drum is normally made of food grade stainless steel.

3.7 Metal of Bamboo Steamer

Steaming baskets are frequently used during Chinese cooking. They are used for steaming all kinds of food, including dumplings, fish, leavened breads, meat, and vegetables.

3.8 Cooking Chopsticks

Very long, large chopsticks, usually about 30 or 40 centimetres (12 or 16 inches), are used for cooking, especially for deep frying foods.

121

Part Two

3.9 Wok Brush

A wok brush is made of a bundle of split bamboo stalks that is tied at the top. It is used to clean woks with the wok still hot and with hot water or water added to a heated wok. It is believed to be a natural and effective way to clean a wok without using chemicals.

3.10 Chinese Rolling Pins

Used to roll out disks of pastry for dumplings etc. Other styles are used as well to chamfer the edges of the pastry. They are thin at one end and thicker at the other end.

Part Three: Chinese Culinary Techniques

Chapter 8 Fundamental Knife Skills and Cutting Techniques

Cutting techniques refer to the methods of using knives to cut ingredients into certain shapes or forms. Chinese cutting techniques have been created and passed on through the practice of generations of Chinese chefs over a long period of time. It would continue to improve as a result of the constant advancement of cooking techniques.

Chinese knife skills emphasize precision, speed, creativity, and aesthetic pleasure. Chefs who utterly understand and master cutting techniques would be able to develop knife skills to a high level.

Section 1 Introducing Chinese Knives (by function)

There are a good variety of knives used in Chinese culinary practices including slicing knife (thin knife), cutting knife, cleaver (chopping knife), cutting and chopping knife, roast duck slicing knife, mutton slicing knife, filling knife, scissors, tweezers and scraping knife, and carving knives.

Part Three

1.1 Slicing Knife

Slicing knife is made of high-quality stainless steel. It is light weighted with a narrow and thin body and a blunt blade. The slicing knife is easy to use and primarily used to cut ingredient into slices, batons, and shreds.

1.2 Cutting Knife

With a broader body, a cutting knife has a wider application as well. It can be used to create slices, batons, shreds, cubes, and blocks. It is also good at cutting ingredients with a firmer texture or meat with small bones. The cutting knife is one of the most used knives in Chinese cooking.

1.3 Cleaver (Chopping Knife)

A cleaver, or chopping knife has a broad spine hence heavier in weight. Comes in various sizes with a rectangular blade, it is usually heavy and is traditionally used for chopping through ingredients with bones or of a firm texture such as pig's head, ribs, or trotters.

1.4 Cutting-Chopping Knife

With an average weight of 750g, a cutting-chopping knife is similar to a cutting knife in size but thicker on the top part of the knife. The front part of the blade is thin and sharp. This knife has the features of a cutting knife and a chop knife.

1.5 Roast-Duck Knife (Small Slicing Knife)

A roast-duck knife is slightly narrower and shorter than a slicing knife. It is lightweight with a sharp blade. A roast-duck knife is mostly used to slice a roasted duck.

1.6 Kitchen Scissors

A pair of kitchen scissors is considered as an accessory tool in the knife family. It is mostly used to aid the preparation of fish, shrimps and vegetables.

1.7 Carving Knives

With a great variety in types, carving knives are specially designed for food carving. Chefs may design their own carving tools to meet their specific needs.

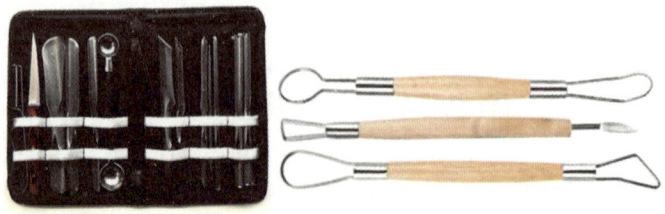

Starting from the left to the right, they are the carving knife (the 2nd from the left), U-shape poker (the 3rd, 4th, and 5th from the left), V-shape poker (tools in the right-hand side pocket), baller (the 6th from the left), score carving tools (see the picture on the right), and module cutters.

- Carving knife, also known as the master knife, is the most used tool for scoring, carving, slicing, screwing, scraping, and hollowing.
- U-shape poker and V-shape poker come in different sizes. The correct posture should be to grasp the ingredient with one dominant hand, and use the thumb, middle finger, and index finger of the other hand to hold the poker at 3 cm above the tip. Poke the tip gently into the ingredient. The ring finger should push against the ingredient to control the force.
- Score carving tools. It is used to carve mountains, stones, crease of garments, and the bone structure of animals.
- Baller is for making balls out of melons of various kinds.
- Module cutters in a range of shapes.
- Other tools including tweezers and uneatable glue.

Section 2 The importance of Cutting Techniques in Chinese Cooking

Simply put, cutting technique is about using appropriate knife to follow a variety of methods to cut ingredients (uncooked, par-cooked, or cooked) into desired sizes and shapes as per recipes. A chef's ability to apply various cutting techniques to enhance the presentation, flavour, or texture of dishes is essential to a Chinese cuisine chef, indicating his/her technical levels.

2.1 Functions of Cutting Techniques in Chinese Cooking

1. To Accelerate Cooking Process
   To cut ingredients into slices, chunks, dices, shreds, batons, cubes, brunoises, fine brunoises of uniform shape and size is a basic requirement in Chinese cooking. Evenly cut items require a shorter time to be cooked evenly.
2. To Increase Blending of Flavours
   To cut ingredients into smaller pieces or scored on the surface would help the flavour of seasonings penetrate to increase flavour blending so that the taste of the cooked food will be even.
3. To Make Bite-size pieces for easy eating
   Chopsticks and soup spoon are used to enjoy Chinese food. To cut ingredients into bite-size pieces would make the food easy to eat.
4. To Enhance Food Presentation
   Basic knife skills can beautify ingredients by creating shapes of uniform specifications. More advanced knife skills are able to showcase great aesthetic pleasure. For example, scoring on the surface of certain ingredients in a pattern, once heat up, the ingredients would form various artistic shapes. Ingredients could also be cut into forms of golden fish, bird, butterfly, plum blossoms etc. to enhance food presentation.

2.2 Requirements of Applying Cutting Techniques

1. The correct body postures
   Stand steadily with your upper body slightly forward, keep a 10cm-distance from the chopping board, keep your chest slightly open in a relax manner. Chopping board should be placed at an appropriate height so you do not have to hunch over the chopping board. Keep both of your eyes on your hands when cutting ingredients.

Part Three

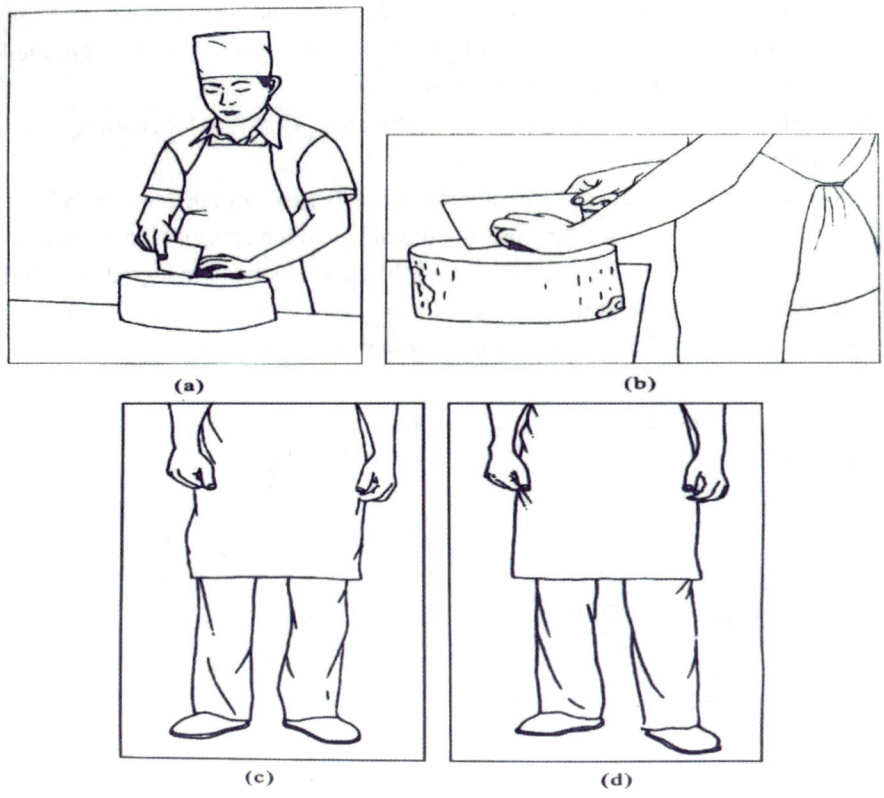

2. Holding the knife
   The way you hold the knife will be determined in part by the cutting technique you use. For example, for straight cutting, you should hold the handle with three fingers and grasp the blade firmly between the pad of your thumb and the side of your index finger just in front of the bolster as shown in the picture.

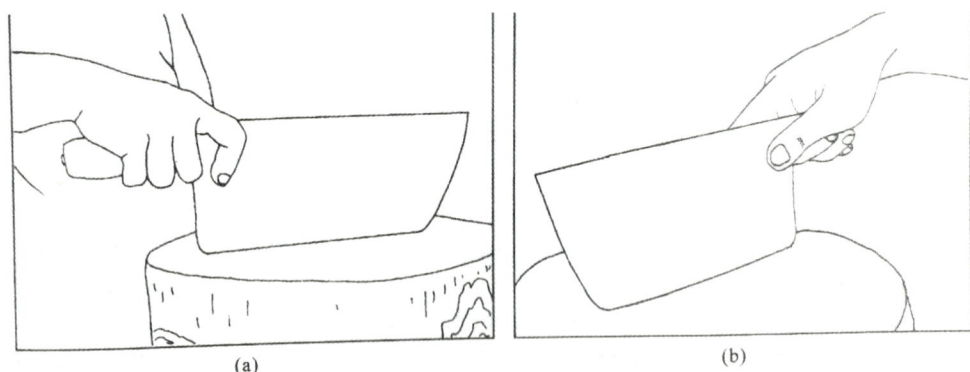

The hand not holding the knife is known as the guiding hand; it is crucial to be aware of the right position of your guiding hand. The fingertips should be tucked under and resting lightly on the ingredient, with the knuckles slightly forward. Remember to keep your thumb tucked behind the fingertips. The knife blade then rests against the knuckle of the middle finger, preventing the fingers from being cut.

3. Mental Concentration
   When cutting, concentrate on your task at hand. Resist the temptation to look around, or let your mind drift away, to prevent injuries from happening.

4. Use the right techniques for the task

   The choice of cutting techniques should be made based on the characteristics of the ingredients of a dish and the cooking method(s) to make it.

   For example, explosive stir-frying and standard stir-frying dishes require ingredients to be cut into smaller and thinner pieces; whilst larger and thicker chunks are suitable for stewing and braising dishes considering their much longer cooking time. Some dishes are particularly made to showcase aesthetic pleasure, it would be appropriate to use scoring techniques.

   When considering the characteristics of ingredients, there are some principles to follow in cutting beef, pork, chicken, and fish into shreds.

   Beef has a tougher texture and plenty of tendons in the meat. It should be sliced against the grain to break the tendons so that the beef shreds would taste tenderer once cooked.

   Pork, on the contrary, is tender without many tendons. It should be diagonally sliced against the grain. By doing so, the pork shreds would not break easily during the cooking and it will remain the tender texture once cooked.

   Comparing to beef and pork, chicken meat is very tender and smooth with little tendons. To get thin and long chicken shreds, the meat should be cut along the grains. If you slice the meat against or obliquely against the grain, you will end up with shreds that break up easily.

   Fish meat is not only tender and smooth but also contains a high level of moist. When slicing, make sure it is cut along the grains into larger size shreds.

5. Ingredients should be cut into pieces of uniform shapes and sizes to allow even cooking.
6. Once finish with cutting, the treated ingredients should meet the desired specifications in terms of sizes and shapes.
7. Fully utilise ingredients. When performing cutting techniques, you should think carefully how to make the most use of the ingredients to avoid wasting.

Section 3 Cutting Techniques

3.1 Straight Cutting Technique

Straight cutting refers to a group of cutting methods in which the blade is at right angles to the surface of the cutting board or the ingredients.

3.1.1 Cutting

Straight-cutting, or Hop-cutting

Method

Keep blade vertical to the cutting board when cutting, move the knife upward and downward in rhythm. The force is placed on the front part of the knife.

This method is suitable for cutting crunchy vegetables such as fresh bamboo shoots, Chinese lettuce.

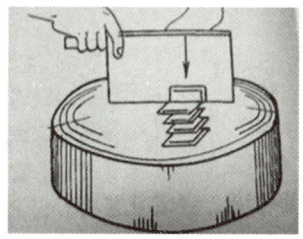

Key Knife Skills:
- Keep the blade at right angles to chopping board.
- Both hands must move in harmony and in rhythm. Using the guiding hand to hold ingredient steadily while moving it backward after each cut; at the same time, cut the ingredient straight down with the hand that holds the knife and move it forward after each cut. The thickness or length of each cut must be uniform. Your ability to move your guiding hand in rhythm to form even intervals during the cutting process is the key to keep the cuts even.

Part Three

- The fingertips should be tucked under and resting lightly on the ingredient, with the knuckles slightly forward. Remember to keep your thumb tucked behind the fingertips. The knife blade then rests against the knuckle of the middle finger, preventing the fingers from being cut.

Push-cutting

Method

Keep blade vertical to the cutting board and ingredient, push the knife down and slice forward to cut the ingredient. The force is placed on the back part of the blade.

This method is suitable for cutting ingredients with a tender and delicate texture or firm ingredients but thinner and smaller in sizes for instance dried tofu, Chinese mustard green, livers, kidneys, meat shreds, meat slices, tripes.

Key Knife Skills:

- Push knife straight down and forward with the right amount of force to provide clean cuts. Each slice only needs one cut.
- The amount of force used in cutting depends on the texture of the ingredients. Ingredients with a tender texture, such as livers and kidneys, are best to be cut gently, while ingredients that are more substantial, such as Chinese mustard green, cured meat, and tripes, will need to be cut slowly.

Pull-cutting, or Drag-cutting

Method

Keep blade vertical to the cutting board and ingredient, push the knife down and slice backward to cut the ingredient. The force is placed on the front part of blade edge.

This method is suitable for cutting boned substantial animal ingredients such as poultry, fish, and meat.

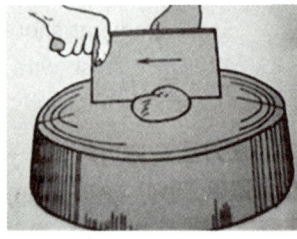

Key Knife Skills:

- The blade rests against the knuckle of your middle finger of the guiding hand. Pull the knife blade backwards in a steady manner. Each slice only needs one cut.
- When pulling the knife, tilt the heel of the blade slightly up and slicing backwards. The force is place on the front part of the cutting edge. When almost cut the ingredient through, push and slice forward with the blade briskly to finish it off. the final step looks like push-cutting but actually it is used to finish off the pull-cutting.

Saw-cutting, or Push-pull-cutting

Method

Push the blade forward first and then pull it backwards to slice like a saw.

This method is suitable for slicing ingredients with a loose texture which would be difficult to keep its shape if using straight cutting, pushing cutting, or pull cutting. For instance, mutton rolls for hotpot, par-cooked pork belly, ham, and bread.

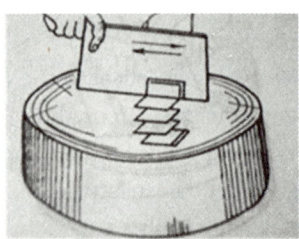

Key Knife Skills:

- Keep the blade at right angles to chopping board when cutting.
- At the beginning of cutting, use the blade in a sawing motion slowing and gently to cut halfway (or 2/3) into the ingredient, then saw-cut with force.
- Make sure that the guiding hand holds the ingredient in place firmly.

Chop-cutting
Method
Chop cutting can be done in two different ways.

1. The food to be chopped should be placed on a firm cutting surface. Grasp the knife firmly, place your guiding hand on the front part of the blade's spin, push the tip down to the chopping board, hold it in place, then tilt the heel of the knife up aiming at where you would like to chop through, and then hit down with force.

2. Grasp the knife firmly with one hand, place the knife on top of the food to be cut, place your guiding hand on the front part of the blade's spin, push the knife down with one hand and then the other, carry on with this motion until the food is cut through.
   Alternatively, place the cutting edge of the knife on the food to be cut, hit the blade's spin with the palm of your guiding hand to finish off the blow.

This method is suitable for cutting ingredients with shells or soft bones, for example, carbs, duck heads; or small and round food such as nuts, cooked eggs etc.

Key Knife Skills:
- When using the first method, it is important to keep the food item still in position and make sure that the knife is aiming correctly at where you would like the food to be cut. The speed of cutting is a key.
- When using the second method, in addition to what mentioned above, make sure both hands push the blade down with equal force.

Roll-cutting
Method
Keep the blade at right angles to chopping board. Begin by making one diagonal cut at one end of the ingredient, then roll the ingredient to one direction with the guiding hand, and make the next diagonal slice. Continue in this way until you have chopped the entire ingredient into evenly sized, diamond-shaped chunks.

This method is suitable for slicing large vegetables in cylinder shape or egg shape such as courgettes, Chinese white radish (mooli), and potatoes.

Key Knife Skills:
- Control the rolling of the ingredient with the guiding hand. The angle of rolling is determined by dish specifications. The wider the angle is, the bigger the piece would be or vice versa.
- Hold the blade so that it is cutting through the food on an angle; the wider the angle, the more elongated the cut surface will be. The smaller the angle is, the wider the cut surface area will be.

3.1.2 Chopping/Mincing
Chopping is a technique for mincing ingredients to make fillings and balls. Chefs have the choice to use one knife or two knives for the task. Chopping rapidly with two knives in unison leads to fast results.

Chopping with Two Knives
Method
Chopping with two knives is a fine-chopping technique. Keep both knives at right angles to the chopping board. The head of two knives should be 5-7cm apart while the distance between the heal of two knives is wider, about 8-10cm. For rapid chopping, move one knife up and one knife down, chop the food from the left to the right

Part Three

and then from the right to the left until it is rather spread out over the chopping board. Scrape it into a pile and chop again. Repeat this process until the food reaches the desired state.

This method is suitable for mincing tender and boneless ingredients.

Key Knife Skills:

- It is important to be able to synchronise the movements of two hands.
- Be aware of keeping two knives apart while chopping. Remember to scrape the food being chopped into a pile when it is spread out over the chopping board from time to time.

Chopping with One Knife, or Straight chopping

Method

Hold the food with the guiding hand, chop through precisely at where it intended to be with one blow for a clean and neat result.

This method is the best for chopping ingredient with bones and with a relatively firm texture into smaller pieces.

Key Knife Skills:

- For a better result, ingredients should be sliced and diced before chopping.
- Wet the knives to prevent food from popping up or sticking to the knives when being chopped.
- Control the force used for chopping to avoid damaging the surface of the chopping board.

3.1.3 Hacking

Hacking is a technique for cutting ingredients with bones or with a tough texture into smaller pieces. It requires to grasp the knife tight and chop with force. There is the straight hacking method and the repeated hacking method.

Straight Hacking

Method

Hold the ingredient to be cut with your guiding hand, aim at the targeted place, raise the knife higher than your head, chop thought with one big blow.

This method is suitable for lager animal ingredients with bones such as port ribs, whole chickens, whole ducks, big fish heads etc.

Key Knife Skills:

- This method requires accuracy, fast speed, and good arm strength.
- The ingredient to be cut should be place on steadily on a flat surface. If your guiding hand is required to hold the food, make sure it is positioned at a safe distance from where the knife lands to avoid injuries.
- Hold the knife tight while hacking. It's the best to cut through the food with one big blow only at each intended place.

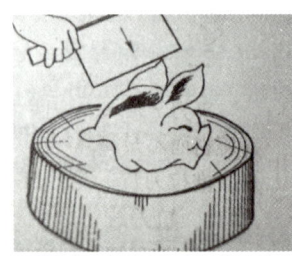

Repeated Hacking

Method

Embed knife into the food at where it is intended to cut with one straight hack, raise the knife high with the ingredient on, drop the knife down with force to the chopping board, repeat this movement until the food is cut through.

This method is suitable for cutting large animal ingredients in a round shape with big bones or with a tough texture such as pig's heads, big fish heads, or shanks

etc. Cutting these types of ingredients normally requires more than one blow.
Key Knife Skills:
- Make sure the knife is embedded in the food firmly. If necessary, use your guiding hand to hold the food when moving up and down.

3.2 Flat Slicing Technique

This is a technique for splitting food horizontally into thinner pieces while retaining its overall shape. When flat slicing, the blade of the knife is parallel to the chopping board.

3.2.1 Push-slicing
Method
Hold the blade of the knife parallel to the chopping board. Place your guiding hand on top of the piece of food to keep it steady. Slice sideways into the food and push the blade away from your body.
This method is suitable for cutting crunchy vegetable ingredients into thin flat slices. For example, Chinese leaves, bamboo shorts, pickled mustard greens etc.
Key Knife Skills:
- Place your guiding hand gently on top of the piece of food to keep it steady.
- Raise the fingers of your guiding hand slightly when the blade passes through underneath.
- Separate the middle finger and the index finger of your guiding hand so that you can observe the thickness of the flat slice.

3.2.2 Pull-slicing
Method
Hold the blade of the knife parallel to the chopping board. Place your guiding hand on top of the ingredient to keep it steady. Slice sideways into the ingredient and pull the blade toward your body.
This method is suitable for cutting animal ingredients with a firm texture into thin flat slices. For example, chicken, pork etc.
Key Knife Skills:

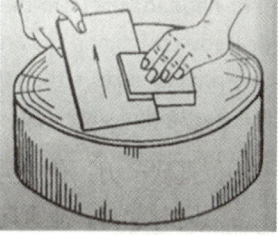

- Hold the knife tight and steady, always keep the blade parallel to the ingredient to ensure uniform thickness of the cuts.
- Separate the middle finger and the index finger of your guiding hand so that you can observe the thickness of the slice. Raise the fingers of your guiding hand slightly when the blade passes through underneath.

3.2.3 Push-pull-slicing
Method
This method is a combination of push slicing and pull slicing like sawing a slice horizontally. You can cut the first slice from the top or the bottom of the ingredient. If starting from the top, it is easier to control the thickness of the cuts. Starting from the bottom is easier to remain the shape.
This method is suitable for cutting large animal ingredients without bones and with a firm texture such as pork, ham etc.
Key Knife Skills:
- Hold the knife tight and steady, always keep the blade parallel to the chopping board/ingredient to ensure uniform thickness of the cuts.
- When start slicing from the top of the ingredient, gently press the ingredient with your guiding hand

fingers to keep it steady, separate the middle finger and the index finger to observe the thickness of the flat slice.
- When start the task from the bottom of the ingredient, use the palm of your guiding hand to stabilise the ingredient, observe the distance between the blade and the chopping board to determine the thickness of the slices.
- To ensue slices with an even thickness, keep the blade parallel to the chopping board throughout the cutting process.

3.2.4 Straight slicing
Method
Keep the blade parallel to the chopping board, cut sideway into the food, straight through.
It is suitable for horizontal slicing ingredients with a tender texture and without bones such as tofu, jelly, pig blood curd.
Key Knife Skills:
- Use a sharp knife to gently cut sideway into the food, then straight through.

3.3 Oblique Slicing Technique

Oblique slicing technique includes two slicing methods in which the blade is kept at angles that are less than 90 degree.

3.3.1 Oblique slicing
Method
Press the ingredient with your guiding hand fingers to keep it steady. If your right-handed, cut from the top left into the surface of the ingredient with an angle, move the blade towards the bottom left direction. The diagonal sliced pieces have a lager surface than that of the straight cut pieces.

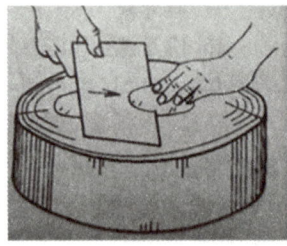

This method is suitable for slicing ingredients with a tender and crunchy texture or small-sized ingredients with a firm texture. For example, chicken, meat, kidney, fish, tripe, and Chinese leaves etc.
Key Knife Skills:
- Use a sharp knife to gently cut sideway into the food, then straight through

3.3.2 Backhand oblique slicing
Method
As shown in the picture, cut from the top into the surface of the ingredient at a slight angle, move the blade forward and away from your body.
This method is suitable for ingredients with a crunchy texture, such as Water bamboo, Chinese leaves, Celtuce (stem lecture), cucumbers, Chinese white radish (moolis).

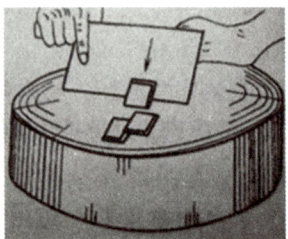

Key Knife Skills:
- The knuckle of the middle finger of your guiding hand should be used as the guide for cutting.

3.4 Scoring Technique

In Chinese culinary terms, scoring means to cut patterned slits on the surface of a piece of food. The depth of the cuts and the distance between each cut should be uniform. This process will help to get ingredients cooked faster and more evenly. It also gives food an attractive appearance once it is cooked.

Chinese Culinary Techniques

This technique is mainly used on ingredients with a firm but crunchy texture, for instance, kidneys, tripes, livers, gizzards, squids, cuttlefish, and whole fish etc.

Scoring technique is an advanced knife skill. There are three common scoring methods, namely straight scoring, oblique scoring, and reverse oblique scoring.

3.4.1 Straight scoring
Method

Straight scoring is similar to straight cutting except that it doesn't require to cut through the ingredient.

It is suitable for processing vegetable ingredients with a crunchy texture and animal ingredients with a firm texture such as cucumbers, kidneys, gizzards, cuttlefish etc.

Key Knife Skills:
- The depth of a cut depends on the texture of the ingredient. Averagely, the depth is between ½ to ¾ of the totally thickness of the ingredient.
- An appropriate angle at which the cut was made and an even distance between each cut are the key to create a beautiful pattern.

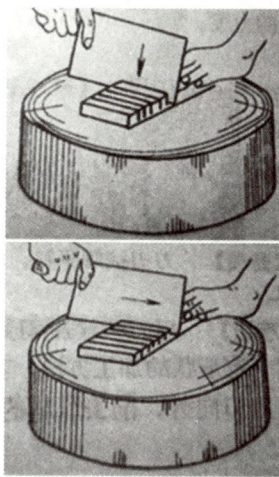

3.4.2 Oblique scoring
Method

Oblique scoring is similar to oblique slicing except that it doesn't require to cut through the ingredient.

It is suitable for processing ingredients with a firm texture cuttlefish, whole fish etc. Combined with other cutting techniques, it could also create special forms such as squirrel shape and grape shape.

Key Knife Skills:
- The depth of each cut must uniform.
- The distance between each cut must even.

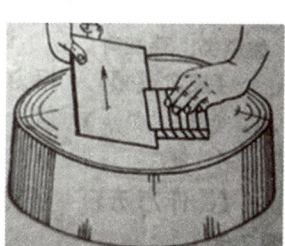

3.4.3 Backhand oblique scoring
Method

Backhand oblique scoring is similar to backhand oblique slicing except that it doesn't require to cut through the ingredient.

It is suitable for processing ingredients with a firm such as squids, pig's kidneys, whole fish etc. Combined with other cutting techniques, it could also create special forms such as wheat ear shape and eyebrow shape.

Key Knife Skills:
- Knife skills should be focused on creating an appropriate angle at which the cuts was made, an equal distance between each cut, and an even depth (between ½ to ¾ of the totally thickness of the ingredient) of the cuts.

3.5 Other Cutting Techniques

3.5.1 Scraping
Scraping by knife is a method to remove the outer skin of ingredients or unwanted part from the surface of ingredients. For example, fish scales, or dirty spots on a carrot. It does not involve cutting.

Grasp the knife with your dominate hand, use the guiding hand to hold one end of the ingredient. Scrap along the upside surface, moving from your guiding hand to the other end. Make long, even strokes. Apply a little pressure, but not much.

3.5.2 Paring
Paring is used to cut off a layer of the outer skin of ingredients or trim ingredients into certain shapes with a knife. Be aware of not to pare too thick layer off.

3.5.3 Hammering
Hammering is to use the spin of the knife to smash ingredients. Scrape the ingredients into a pile from time to time during the process. It is suitable for processing tender animal ingredients such as fish meat, chicken breast. This process is similar to fine chopping except that it uses the spin not the cutting edge of the knife.

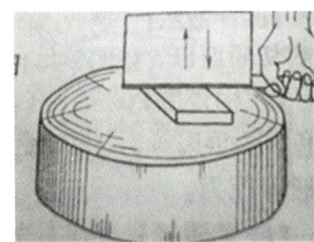

3.5.4 Smashing
Smashing is to use the side of the knife to hammer ingredients to either loosen them up or break them down into pieces. It is mostly applied to process gingers and spring onions helping to release their aroma. There is a popular side salad dish called "Smashed cucumber" in which both cucumber and garlic are prepared using this method. It also can be used to tenderise pork chops and beef steaks.

3.5.5 Deboning
It is used to separate parts or remove the bones from meat, poultry, or fish. The deboning skills include using different parts of the knife (such as tip, the heel, the cutting edge) to remove bone while remain the overall shape of ingredients.

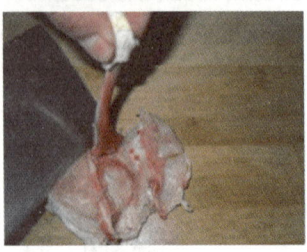

3.5.6 Mashing
Using a knife to mash tender ingredients such as cooked potatoes or tofu involves pressing and pulling with the knife blade until reaching a fine consistency.

Chinese Culinary Techniques

Section 4 Knife Skills

By applying various Chinese cutting techniques, ingredients can be processed into chunks, slices, shred, batons, dices, mince, paste, and mash.

4.1 Chunks

An illustration of the cut shape	Dimensions L/W/H	Method	Suitable ingredients
Elephant eyes (Rhombus)	3cm x 3cm x1.5cm	Cut the ingredient into slices to the thickness that you wish the finished chunks to be, stack the slices on top of one another and make even parallel cuts to the appropriate thickness, gather the barons together, make parallel cuts on an angle through the batons. Chopping or hacking may be applied, depends on the texture of the ingredient	Tender or crunchy vegetable ingredients or boned animal ingredients that are tender
Big Squares	Larger than 3.3cm x 3.3cm x 3.3cm	Cut the ingredient to slices to the thickness that you wish the finished squares to be, stack the slices on top of one another and make even parallel cuts to the appropriate thickness, gather the barons together, make parallel cuts through the batons. Chopping may be applied, depends on the texture of the ingredient	Tender or crunchy vegetable ingredients or boned animal ingredients that are tender
Small Squares	3.3cm x 3.3cm x 3.3cm or under		
Rectangles	Rectangular cubes that are larger than 2.5cmX 2cmX 1.5cm		
Dominos	Rectangle cubes that are smaller than 2.5cmX 2cmX 1.5cm		
Wood logs	4.5cm x 1.5cm x 1.5cm	Halve the ingredient lengthwise, smashing one piece at a time with one blow to loosen it up, then cut it into irregular thick batons. The shape is similar to wood logs hence the name.	Water bamboo and winter bamboo or other similar textured vegetable ingredients. Cucumber salad
Roll cuts	The bottom of the edges should be no more than 3cm wide	Cutting through the ingredient diagonally, rolling it inwards after each cut and slice through the ingredient on the same diagonal, forming a piece with two angled edges	Long, cylindrical or egg-shaped vegetables such as cucumbers, potatoes, Chinese yams, parsnips, carrots, and Chinese lettuce (celtuce)

Tips: cooking methods and the characteristics of ingredients determine the choice of chunk shapes. For example, larger chunks are the best for dishes that are cooked over a low heat for a long period of time whilst smaller chunks are suitable for dishes that are cooked over a high heat for a short period of time. For tender or crunchy ingredients, it would be appropriate to process them into bigger chunks. For tough ingredients with bones, smaller chunks would be a better choice.

4.2 Slices

An illustration of the cut shape	Dimensions Length x Width X Thickness	Method	Suitable ingredients
Willow leaves	6cm long and 0.3cm thick	Cut in half the ingredient diagonally, then slice each piece with the oblique slicing technique	Pork liver or similar ingredients
Elephant eyes (Rhomboid)	5cm x 2.5cm x 0.3cm	Apply straight cutting techniques to cut he food into elephant eye shaped chunks first, and then cut into slices	Crunchy and tender vegetables
Crescents (wedges)	The length is no longer than 4cm	Halve the ingredient lengthwise from the centre, place each half face down on the cutting board, repeat this process with each cut piece to create wedges in desired sizes	Round or egg-shaped ingredients
Half-splits	Rectangular or disk-shaped slices. The thickness of each slice is between 0.3-1cm	Each finished half-split slice requires two cuts. Firstly, make a cut that is more than a halfway into the ingredient on right angles, then make a parallel cut to cut the piece off to complete the half-split	Tender ingredients such as aubergine slices, pork slices
Rectangles	6-6.6cm x 2-3cm x 0.3-0.5	Cut the ingredients into rectangular chunks, and then cut them crosswise into slices	Vegetable, meat, and poultry ingredients
Dominoes	4.5-5cm x 1.6-2cm x 0.3-0.5	Cut the ingredients into rectangular chunks, and then cut them crosswise into slices	Vegetable, meat, and poultry ingredients

Tips: grip the knife with your dominant hand firmly and steadily, gently press on the ingredient with your guiding hand to keep it stable on the chopping board, wipe clean the surface of the chopping board and your knife whenever it is needed while using them.

4.3 Shreds

An illustration of the cut shape	Dimensions	Method	Suitable ingredients
Depending on the thickness of the shreds, shreds can be further divided into silver needle shreds, hemp thread shreds, matchstick shreds, incense stick shreds	The length of a shred is between 5cm and 8cm	Straight cut or flat slice the ingredients into slices, and then either • pile the slices up by placing one on top of another, aligning the edges, and make parallel cuts of the same thickness through the pile. Or, • partly stack one slice on top of another, like the way that roof tiles are laid. Or, • for thin and large ingredients such as lettuce and omelette, roll them up into a cylinder, and make parallel cuts of the same thickness through the roll.	Vegetable, boned meat, and boned poultry ingredients

Chinese Culinary Techniques

Tips:
1. Regardless of the type of stacking method, make sure that the pile of slices is not too tall.
2. Stabilise the ingredient with your guiding hand to enable parallel and even cuts.
3. Remember that tough meat like beef should be shredded against its grain to break down fibre; tenderer meat such as pork should be shredded along or obliquely against its grain to make stronger shreds that would not break up easily during the cooking; chicken and pork tenderloin are very tender so they must be shredded along the grain, otherwise the shreds will break up during the cooking.

4.4 Strips

An illustration of the cut shape	Thickness	Method	Suitable ingredients
Thin strips	1cm	Cut the ingredient into strips in desired thickness.	Vegetables and boned meat and poultry
Thick strips	3.5-4.5cm		

4.5 Dice, Fine Dice, Mince

An illustration of the pattern	Dimensions	Method	Suitable ingredients
Dice	1.5cm x 1.5cm x 1.5cm	Dices are made by cutting the ingredient into slices to the thickness that you wish the finished dice to be, stack slices up with edges aligned and make even parallel cuts lengthwise into strips. Gather the sticks together; using your guiding hand to hold them in place, and make crosswise parallel cuts through the strips. The finished dices should have uniform dimensions	Vegetables and boned meat and poultry
Fine dice	0.8cm x 0.8cm x 0.8cm		
Mince	In similar sizes of a millet grain		

4.6 Mash and Paste

An illustration of the pattern	Method	Suitable ingredients
Mash and Paste	Remove bones, tendons, and skin, cut the ingredient into slices, then into shreds, then into small dices, and finally chop with two knives until reach the desired consistency. To make fish/prawn mash/paste, add a piece of pork fat to enhance the stickiness	Chickens, prawns, fish, and meat

Part Three

4.7 Patterns

An illustration of the pattern	Method
Wheat Ear	1. Place a square or rectangle piece of ingredient diagonally on the chopping board, use oblique scoring technique to diagonally penetrates 4/5 of the total thickness against the grain, make parallel cuts about 2mm apart throughout. 2. Straighten the ingredient, using the oblique scoring technique to cut crosswise at a 45-degree angle about 4/5 deep along the grain, make parallel cuts about 2mm apart throughout. 3. Finally, cut the scored ingredient into 5cm x 2.5cm pieces Once heated up, each piece would roll up to form a pattern that looks like wheat ear
Lychee	1. Prepare an 8mm-thick piece of ingredient, use backhand oblique scoring method, cut crosswise approximately 2/3 into the food against the grain, make parallel cuts at every 5mm throughout 2. Turn the ingredient 90 degree to either side, repeat the step 1 3. Finally, cut the scored ingredient into 5cm x 3cm rectangles, rhomboids, or triangles Once heated up, each piece would shrink and curled to form a pattern that looks like lychee
Chrysanthemum	1. Place a rhombus-shaped piece of ingredient diagonally on the chopping board, use oblique scoring technique to diagonally cut into 4/5 of the ingredient against the grain, make parallel cuts throughout to make thin slices. 2. Turn the ingredient 90 degree to either side, repeat the step 1. At this point, the piece should look like a bundle of shreds with one end connected When heated up, the piece would shrink and curled to form a pattern that resembles chrysanthemum flower
Roll	1. Use straight scoring technique, straight cut 4/5 of the total thickness into the ingredient against the grain, make a parallel cut at every 2mm throughout 2. Turn the ingredient 90 degree to either side, repeat the step 1 3. Finally, cut the scored ingredient into 5cm x 3cm rectangular. Once heated up, each piece would shrink and roll up

The above scoring techniques are suitable to apply on animal ingredients with parallel muscle that has no fibre, for example squid, black carp, cuttlefish, mandarin fish, and pig's kidney etc. These patterns are commonly used to process ingredients for stir frying and Bao (a variation of stir frying, literally means explosive stir-frying).

Chapter 9 Chinese Cooking Techniques

Cooking method refers to the process of turning prepared ingredients into different flavoured dishes through heating and seasoning. With its long history, large territory, varied landscapes, and ethnically diversified population, ingredients and preferred flavours differ from one region to another in China. Through over 2,000 years of practice and innovation, Chinese chefs have created a variety of cooking methods.

Section 1 Chinese Hot Dish Cooking Techniques

Based on the type of heat transfer medium is used, Chinese hot dish cooking methods can be divided into 6 types:

Table 9 Chinese Hot Dish Cooking Methods

Types of Heat Transfer medium	Cooking Methods		
	In Chinese Characters	In Pin Yin	In English
Oil	炸	Zha	Deep frying
	煎	Jian	Pan frying, food would be fried on both sides
	炒	Chao	Stir frying
	贴	Tie	Pan frying, food would be fried on the bottom only
	爆	Bao	Explosive stir frying
	烹	Peng	A variation of stir frying.
	熘	Liu	Frying and then mixed or topped with gravy
	塌	Ta	A combination of pan frying and braising
Water	汆	Cuan	A variation of boiling
	烩	Hui	A variation of stewing
	煮	Zhu	A variation of boiling
	熬	Ao	A variation of stewing
	炖	Dùn	A variation if stewing
	煨	Wei	A variation of stewing
	涮	Shuan	Chinese Hot Pot, or Chinese fondue
	烧	Shāo	A variation of stewing
	扒	Pa	A variation of braising
	焖	Menn	A variation of braising
Steam	蒸	Zheng	Steaming

Part Three

Solid materials	盐焗	Yan Ju	Salt baked
	石烹	Shi Peng	Stone cooked
Radiant	烤	Kao	Grilling, Roasting
	熏	Xun	Smoked
Special medium	蜜汁	Mi Zhi	Honey or sugar liquid coating
	挂霜	Gua Shuang	Sugar marinated
	拔丝	Basi	Caramelised

1.1 Zha (Deep Frying) Techniques

Deep-frying technique is widely applied in Chinese cuisine. Through contact with sizzling hot oil, raw ingredients are cooked quickly. Their crust turns amber and crunchy, while the interior remains tender.

Based on the flavour created, deep frying techniques can be divided into plain frying, dry frying, crispy frying, crunchy frying, soft frying, tender frying, and fragrant and crispy frying.

Deep-frying is not just appreciated as a standalone cooking technique. On many occasions, ingredients are par-fried before finished in other methods.

1.1.1 Qing Zha (Plain Frying)

Qing Zha literally translated as plain frying, which means frying without coating. Ingredients are cut, marinated and then fried without coating. Some items can be part cooked using other methods.

Notes
- It is not recommended to use sugar and dark-coloured seasonings to marinate the ingredients as they may cause food getting darkened in frying.
- If a large or whole sized food needs to be cut into bite-size pieces before serving, it should be done immediately after the frying to maintain its texture and presentation.
- Some of the fried dishes can be served with a dipping sauce.
- Deep frying can be dangerous when using a wok compared to a deep fat fryer, you must ensure you have prepared well in readiness for deep frying.

Key skills
- Ingredients should be cut into similar pieces with uniform sizes for even cooking.
- Ingredients should be marinated evenly before frying (If marinating is required).
- Choice of the heat level and oil temperature should be made based on the texture and shape of ingredients.
- Small sized ingredients should be blanched first at one temperature and then cooked at a higher temperature).
- Large sized ingredients should be fried on high heat first to maintain the shape, and then reduce to medium heat to allow ingredients to be fully cooked, finish with high heat so that ingredients would not absorb too much oil.

Features of plain fried dishes
Thick soup, full-bodied and palatable taste with tempting aroma.

Dishes cooked with this method
Plain Fried Lamb Skewers (Xinjiang Muslin Cuisine)

Plain Fried Pork loin (Shangdong cuisine)

Plain Fried Pork Fillet (Sichuan cuisine)

1.1.2. Gan Zha (Dry frying)

Gan Zha, or dry frying, is a cooking method wherein ingredient are prepared, marinated, dry coated or wet coated before frying. There are 4 dry frying methods:

Powdered dry frying

Marinated ingredient is coated with dry corn starch or flour before frying.

Battered dry frying

Marinated ingredient is dipped in thin batter before frying.

Meat Ball dry frying

Meat Ingredient is blended to paste, then made into meat balls before frying.

Steamed dry frying

Meat ingredient is blended to paste, then steamed (in molds or wrapped to form different shapes), cut into smaller pieces, and finished with frying.

Notes
- Be careful with the thickness of the flour or corn starch coating. If it is too thin, it would not help to keep ingredients tender. If it is too thick, it will result in hard crust.
- If use dry flour or corn starch, it needs to be firmly pressed onto the ingredients to make sure the coating stays.
- Plunge one piece of ingredient into hot oil at a time to avoid pieces get stuck together.

Key skills
- Ingredient is evenly marinated and dry or wet coated before frying.
- Ingredient is dipped in dry flour or corn starch right before frying.
- The dry coating is made evenly all around all. Shake off loose flour or corn starch before frying.
- Prior to fry the ingredients for the second time, food residues are removed for the oil.
- Batter is coated evenly all around the ingredients.
- Coated ingredients are fried over high heat first to allow crust to be formed quickly in order to keep the shape.
- Choose an appropriate cooking time and oil temperature based on the sizes of the ingredients. Frying large sized ingredients on the wrong heat will cause the food being burnt from outside but not thoroughly cooked from inside. Similarly, when frying small sized ingredient, too long a cooking time will result in tough texture.

Features of dry fried dishes

Dry fried dishes are golden in colour, crusty, crispy, dry, and aromatic with a combined flavour of savoury and umami.

Dishes cooked with this method

Dry Fried Prawn meatballs

Dry Fried Prawn Rolls (Cantonese Cuisine)

Dry Fried Chinese tapertail anchovy, or Coilia nasus (Jiangsu Cuisine)

Dry Fried Cuttlefish Roll (Beijing Cuisine)

1.1.3. Ruan Zha (Soft frying)

Ruan Zha, or soft frying, is a technique wherein small sized tender ingredients are marinated first, then dipped in egg white/egg yolk/whole egg batter, and finally fried on medium heat.

Notes
- Peanut oil is a preferred choice for soft frying due to its pure colour and quality which keeps ingredient soft in frying.
- Plunge one piece of ingredient into hot oil at a time to avoid pieces getting stuck together.
- Remove oil completely before serving, sprinkle a splash of sesame oil and Sichuan pepper salt before serving.

Key skills
- Choose tender and less stringy ingredients, which are boned and skinless.
- Cut ingredient into dices, batons, or slices.
- Add seasonings to minced meat and stir it well.
- Wipe off excessive water with paper towel before coating.
- If use egg white batter, dip and fry one piece at a time.
- Medium oil temperature (not higher than 180°C) is required to achieve the desired colour and doneness.
- Soft fried dishes are normally served with condiments such as catch-up, sweet flour paste, or Sichuan peppercorn salt. Condiments can be placed on the side of the plate, or sprinkled on top of the food, or in a small saucer plate.

Features of Soft Fried Dishes

Soft fried dishes are light golden in colour, crispy and soft from the outside and fresh and tender from the inside.

Dishes Cooked with This Method

Soft Fried King Prawns (Beijing Cuisine)

Soft Fried Seasonal Vegetable (Cantonese Cuisine)

Soft Fried Perch Fingers (Zhejiang Cuisine)

Soft Fried Chopped Chicken (Sichuan Cuisine)

1.1.4. Su Zha (Crispy frying)

Su Zha, or crispy frying, is a cooking technique in which ingredient is first boiled or steamed, then dry coated or/ and wet coated (on some occasions coating is not required), and finally fried.

Notes
- Add seasonings to minced meat and stir it well. It is then par cooked by steaming to maintain a tender texture.
- Par-cooked ingredient must be just cooked. It takes a much longer time to fry under par cooked ingredient. Over par cooked ingredient is likely to come apart when frying.
- Herbs and spices such as spring onions, ginger and cassia bark should be removed before coating.
- When seasoning, a balanced flavour should be considered.
- Be careful with the level of heat used when fry Ingredient without coating. Over cooked, it will result in tougher texture.

Key Skills
- Deboning and mincing ingredients.
- Ingredient with bones and unbroken skin should be fried without coating while boned ingredient should be fried with coating. This is a general principle applied to deciding on frying with or without coating.
- Choose tender and fresh animal ingredients for Su Zha (Crispy frying). Large cuts or whole sized in-

gredients are par-cooked by steaming or boiling.
- Choose the appropriate par cooking method(s) and seasoning method.
- Tough and large ingredients are marinated before frying to provide a tenderer texture.
- Small and tender ingredients are boiled with seasonings before frying.
- Steamed and boiled ingredient is easy to break, so, when frying, hold it with a slotted spoon and slide it into oil carefully.
- When batter coating is applied, ingredient is covered evenly with batter before frying.
- 150°C and above oil temperature is required for Shu Zha (Crispy frying) which is hot enough to form crust. Some ingredients may need to fry for the second time. Whole sized food is chopped or sliced immediately after frying and reshaped to reflect its original form when serving.
- Skillful at controlling oil temperature taking into consideration the quantity and size of ingredient, the thickness of coating, the level of heat, and the temperature and amount of oil needed.

Features of Crispy Fried Dishes

Crispy fried dishes are golden or dark golden in colour, extremely crispy, tender, and aromatic.

Dishes Cooked with This Method

Crispy Fried Rump (Shandong Cuisine)

Crispy Fried Lamb (Chinese Islamic Cuisine)

Crispy Fried Duck Breast with Walnut (Huaiyang Cuisine)

1.1.5. Cui Zha (Crunchy Frying)

Cui Zha, or crunchy frying, is a general term for two methods of frying: coated crunchy frying (cui hu zha) and uncoated crunchy frying (cui pi zha).

Notes
- Tender, fresh, moist and boned ingredients such as fish and shellfish are preferred.
- Vegetables, except for fresh mushrooms which is umami, are seldom cooked in this way.
- Do not stir batter too much and never overuse baking powder for it will give an astringent taste to the batter.
- Coated crunchy frying requires two frying phases. The first phase is for cooking and shaping while the second phase is for colouring and getting a crispy shell. Do not break the coat. It will cause oil gets in between the coat and the food.
- Syrup should be brushed evenly on ingredient when it is hot. An unevenly coat will result in uneven colour.
- After brushing with syrup, the food should be dried at a cool and well-ventilated place.
- Oil temperature should be adjusted according to the characteristics of the ingredient and the features of the dishe.

Key skills
- Ingredients are cut into smaller pieces with uniform sizes.
- Ingredient are marinated before frying.
- Making batter with an appropriate consistency.
- Oil temperature is controlled at 90°C - 120°C for initial frying and increased to 150°C - 180°C for the second time frying.
- Plunge ingredient into oil with care to avoid the pieces getting stuck together or breaking the coating or skin of the pieces. Fry ingredients with a larger amount of oil at a lower oil temperature is preferred.

Features of Crunchy Fried Dishes

Crunchy Fried Dishes are golden or jujube red in colour. Umami taste from inside, rich, moist and crunchy from outside.

Dishes Cooked with This Method

Crunchy Skin Chicken (Cantonese Cuisine)

Crunchy Fish Fingers (Jiangsu Cuisine)

1.1.6. Xiang Zha (Fragrant Frying)

Xiang Zha, or fragrant frying, is a cooking method wherein ingredient is marinated, coated with flour or corn starch, dipped in egg and then in breadcrumbs (or steam bun crumbs), or crushed nuts and seeds such as walnut, peanut, macadamia, cashew nut, and pine nut and sesame seed etc., and then finished with frying.

Notes

- Frying with breadcrumbs coating is called "block frying" or "lucky frying".
- When frying sliced ingredient, poke through both sides with the tip of a cleaver for easy infusion. It will also help to avoid the food slices getting curled up with heat.
- Use plain or savoury breadcrumbs, avoid sweet bread as it will darken the colour of the food.

Key skills

- Choose tender and umami meat for Xiang Zha (Fragrant frying).
- Ingredient is cut to long strips, large slices, or fillet. Minced meat is made into balls or burgers.
- Choose appropriate seasonings for marinating ingredient to achieve a balanced flavour. For example, if ingredient tastes fishy, add white pepper powder to remove odor.
- All three coatings are evenly done. When dip in breadcrumbs, press to firm the coat.
- Add additional water to egg to if needed for an appropriate consistency.
- Skillful at controlling the oil temperature with flexibility. In general, 150°C - 180°C is appropriate for Xiang Zha (Fragrant Frying).

Features of Aromatic Frying Dishes

Aromatic Frying dishes are golden, crispy, crumbly, and aromatic.

Dishes Cooked with This Method

Guinea Fowl Fillet

Lucky Prawn Balls

Aromatic Fried Meat Balls

Fish Fillet with Sesame Seeds

1.2 Chao (Stir-frying) Techniques

Chao, or stir frying, is perhaps the most commonly used cooking method in Traditional Chinese culinary arts. The Chinese term "Stir-fry a dish" (Chao Cai) is used as a general reference to catering industry and Chinese cuisine. In northern China, the term "zuo fan" (literally means "making meals") and "chao cai" (literally "stir frying dishes") are used interchangeably. Cooks are referred to as "those who stir fry dishes".

Chao (Stir Frying) cooking method involves cooking bite-sized fresh ingredients with a small to medium amount of oil over medium or high heat.

Chao (Stir Frying) is distinctive in the following ways from other hot dish cooking methods.

- Consumption of oil is kept to a very low level as indicated in some recipes.
- Temperature of oil requested is intense. In general, oil is heated up to between 120°C and 240°C prior to plunge ingredients in.

- Ingredients are often cut into bite-sized dices, strips, shredded, and thin slices.
- The cooking time is short, and the ingredients are rapidly stirred in the wok

There are 5 stir-frying methods, namely Sheng Chao (Stir-frying Raw Ingredients), Shu Chao (Stir-frying cooked ingredients), Gan Chao (Dry Stir-frying), Ruan Chao (Soft Stir-frying), and Hua Chao (Moist Stir-frying).

1.2.1 Sheng Chao (Stir-frying Raw Ingredients)

Sheng Chao can be literally translated as Stir-frying raw ingredients. In this method, pre-cut raw ingredients are cooked in wok with a small amount of hot oil on high heat rapidly.

Notes

- Some dishes are finished by adding corn starch solution as thew thicken agent. The amount of corn starch used is determined by the requirement of a dish.
- Use high heat for a quick stir-frying. Toss the wok from time to time and remove ingredients immediately from wok once they are cooked to avoid getting burnt or stuck on the wok.

Key skills

- Quickly stir fry at constant high heat to maintain the colour and achieve a tender and crunchy texture.
- For dishes with combined ingredients, the timing and sequence to add other ingredients in should be determined by the texture and durability of heat of the ingredients.
- Quick stirring is a key for even heating and efficient infusion.

Features of Sheng Chao Dishes

Dry, aromatic, fresh, and tender are the key characteristics of Sheng Chao dishes.

1.2.2 Shu Chao (Stir-frying par-cooked Ingredients)

Shu Chao, or stir-frying par-cooked ingredients, is a cooking method in which cooked ingredients in the shape of slices, shreds, and strips are stir fried with a small amount of oil. Coating and marinating are not required.

Notes

- Flour paste, fermented bean, or bean paste, if used, must be stir fried until aromatic.
- Some dishes may require adding starch to thicken. It depends on the amount of liquid generated by the heating.
- This method is often used to cook ingredients that cannot be thoroughly stir fry cooked quickly.

Key skills

- Par-cooking skill, more specifically, to what extend the ingredients are cooked during the par-cooking stage should be in line with the characteristics of ingredient and the quality requirements of a dish.
- Selection of seasonings. seasonings with strong aroma and flavour such as chili bean paste, sweet flour paste, and soya bean paste are frequently used to create a balanced savoury and aromatic flavour. Vegetables with strong aroma such as garlic sprouts, garlic stem, shallot and spring onions are commonly used to add more taste to the dish.
- Medium heat is recommended for general use. When a large amount of ingredients involved, high heat should be used. The oil temperature should be controlled at between 150°C to 180°C. Only a small amount of oil is required for cooking, constant stir until ingredients become aromatic, then immediately remove from wok.

Features of Shu Chao dishes

Thick soup, full-bodied and palatable taste with tempting aroma.

Dishes cooked with this method
    Stir Fried Cherry Meat (Shandong cuisine)
    Stir Fried Eels (Henan cuisine)
    Stir Fried Crab Meat (Jiangsu cuisine)
    Stir-Fried Beef Tripe (Beijing cuisine)
    Twice Cooked Pork Slice

1.2.3 Gan Chao (Dry Stir-frying)

Gan Chao, sometime referring to as Gan Bian, is literally translated as dry stir-frying. It is a cooking technique wherein pre-cut ingredients are cooked with a small amount of hot oil on medium to low heat for a relatively long time to reduce moist of the ingredients and to infuse the seasonings into the ingredients.

Notes

- Neither coating (wet or dry) nor thickening is required for dry stir-fried ingredients.
- Wok is the main heat-transfer medium and oil is mainly for smoothing and flavour-enhancing. The oil temperature should be carefully controlled.
- For an enhanced flavour, ingredients should be marinated prior to cooking.

Key skills

- Use a moderate amount of oil. Too much oil will toughen ingredients while too little oil would not be able to remove moist out of the ingredients.
- In case of too much liquid produced in the cooking, continue with stir frying until most of the liquid has evaporated.
- Correct control of the heat level. The general principle is to cook on high heat and then reduce to lower heat. Constant high heat could burn ingredients from outside before water from inside gets evaporated. Equally, only using low heat will create tougher texture.

Features of Dry stir-fried dishes

    Dry stir-fried dish tends to have a dark red colour. It is dry, aromatic, and crispy. Oil is visible in finished dish.

Dishes cooked with this method
    Dry Stir-fried Shredded beef
    Dry Stir-fried Squids
    Dry Stir-fried Bamboo Shoots
    Dry Stir-fried Eel Shreds

1.2.4 Ruan Chao (Soft Stir-frying)

Ruan Chao, or soft stir-frying method involves processing par-cooked ingredients into fluid, paste, or fine dices, mixing them with seasonings, whisked egg, and starch, and then rapidly stir fried on medium to low heat. The mixture could also be oil-blanched first and then soft stir-fried.

Notes

- To make sweet soft stir-fried dishes, sugar and fat should be added when food being cooked releases aroma and becomes hardened. Allow the sugar and fat to melt thoroughly and to completely blend with the food. It will create a sweet aroma and a tender and glossy texture to the dish.
- Once the required texture is achieved, remove from wok immediately to avoid food being browned or producing too much liquid.
- To produce the desired colour and flavour, use plain oil and starch for cooking. Minor ingredients

should be carefully selected so that it enhances the colour and flavour of the dish.

Key skills

- For soft stir frying, ingredients should be in the forms of fluid or paste. Bite-sized boned proteins with a tender texture such as prawns, beef (the tender cuts), chicken liver would be good too.
- To make paste, remove tendons and bones, use mashing technique to mash raw or steamed the main ingredient into paste/mash with a knife. The minor ingredients could be in the shape of small slices or fine dices.
- Ingredients should be mixed with egg, starch and water in readiness for soft stir-frying. The amount of egg, starch and water in the mixture can be adjusted based on: the ingredients' capacity to harden; the quality requirement of the dish.
- In terms of the heat control, pre-heat the wok on high heat, grease the wok and reduce to medium-to-low heat. Plunge in ingredients and quickly stir fry until solidified. If runny ingredients get stuck on wok, pour some vegetable oil on the affected area, continue with stir frying until ingredients hardened. The heat should be moderate.

Features of Soft stir-fried dishes

The liquid in soft stir-fried dishes is half solidified. The taste is savoury and umami with a light, moist, tender, and smooth texture.

Dishes cooked with this method

Osmanthus Egg (Henan Cuisine)

Furong Chicken Slice (Shangdong Cuisine)

Baiyu Chicken Breast (Hebei Cuisine)

1.2.5 Hua Chao (Moist Stir-frying)

Hua Chao, literally translated as Slid-Stir-frying, is a cooking method wherein pre-cut ingredients in the shape of shreds, dices, slices, strips, or granules are wet coated, slide in oil, quickly stir-fried with a small amount of oil on high heat, and finished with thickening by adding corn starch. Wet coating, thickening with corn starch are not always required.

Notes

- Use hands to mix ingredients with seasonings and starch. Hand movement should be gentle and evenly distributed to make sure that all pieces are coated and none of them is broken.
- Corn/potato/sweet potato flour and egg white are commonly used in starching. The consistency of the mixture and how well the mixture sticks onto the surface of ingredients are determined by the characteristics of ingredients and the quality requirements of dishes.
- When stir frying more than one ingredient, it normally starts with stir-frying or blanching in oil, the minor ingredients (referring to ingredients other than the main ingredient) so that all ingredients will be cooked to the same extent.
- When blanching ingredients in oil, heat the wok up prior to pouring oil in, oil temperature is kept between 90°C ~ 150°C. Stir ingredient in oil slowly in one direction with chopsticks and then speed the stirring up to make sure pieces are not stuck together. Remove from wok to drain when ingredients are just cooked.

Key skills

- Being familiar with the texture and properties of ingredients to make reasonable choice. Ingredients for moist stir fry should be fresh, tender, boneless, skinless, and tendon-less.
- Ingredients should be cut into dices, fine dices, shreds, slices, strips with uniform specifications.

Part Three

- Marinating and starching are the key to ensure a tender texture.
- The time used to moist stir fry is short, the speed of cooking is fast. Ingredients are cooked on high heat, so seasonings need to be added in quickly. For a better flow of work, starching mixture should be made prior to cooking. Oil-blanched ingredients should be plunged back in wok immediately and stir fried quickly with the starching mixture. If the above are done slowly, the dish would become tough, chewy, and unsmooth.

Features of Moist stir-fried dishes

Moist stir-fried dishes are tender, soft, smooth, refreshing, moist and rich.

Dishes cooked with this method

Moist stir-fried Chicken breast (Beijing Cuisine)
Beef in oyster sauce (Cantonese Cuisine)
Broccoli and Scallop (Cantonese cuisine)
Shredded Pork with Water Shield (Zhejiang cuisine)

1.3 Liu

Liu, is a method of cooking in which pre-cut ingredients in the shape of slices, dices, shreds or small chunks are cooked first and then mixed or topped with Lu sauce (similar to gravy). Based on the first cooking method used, there are three types of Liu, namely Jiao Liu (Deep-fried Liu), Hua Liu (Moist stir-fried Liu) and Ruan Liu (Soft stir-friedl Liu).

1.3.1 Jiao Liu (Deep-fried Liu)

Jiao Liu, or Deep-frying Liu, is a type of Liu wherein bite-sized ingredients are marinated, batter coated, deep fried in hot oil until crusted, crispy and golden, and then mixed or topped with Lu sauce before serving.

Notes

- The consistency of batter needs to be moderate. Too runny batter will be hard to keep the shape. There is also the danger for batter to crunch up in hot oil which will affect the colour of oil. Equally, too thick batter will result in an uneven coating and tougher texture.
- If dry-coating is applied, make sure flour/cornflour is firmly pressed onto the ingredient so that it won't fall off when frying to affect the colour of oil. When mixing with Lu sauce, the speed of cooking is a key to achieve an aromatic crispy texture.
- The consistency and seasoning of the Lu sauce should be moderate.
- Deep-frying, oil blanching, steaming and boiling are common par-cooking methods used in Liu. Deep-frying and oil blanching are suitable for bite sized ingredients. Large sized ingredients should be prepared with score cutting prior to deep-frying and oil blanching. Steaming or boiling is a better choice of par-cooking method for whole sized ingredients.

Key skills

- Ingredients should be cut to similar sizes for an even heating, easy infusion and good presentation.
- The flavour of the marinade prepared should be savoury. Make sure the ingredients are evenly marinated.
- The consistency of the batter is a vital element. In general, too thick dry coating tends to affect the texture and flavour of the ingredients while too thin a dry coat cannot maintain the moisture and fragrance of ingredients over high oil temperature. Water-based coating should be made slightly thicker to achieve a texture that is crispy from outside and tender from inside.
- Carefully control the oil temperature. For the first time frying, ingredients are cooked on medium

- heat until they become firm. For the second frying, ingredients are fried in intensive heat to quickly become crispy from outside. The oil temperature is usually maintained between 180°C-240°C.
- Proteins such as chicken, fish, shrimp, pork, egg, rabbit meat paste should be shaped and tenderized before steaming on medium heat so that it is easy to cut when cooled down. Ingredients that are dipped in egg, and then cornflour prior to deep-frying should be fried on medium heat (oil temperature is between 150°C-180°C) for a crispy from outside and tender from inside texture.
- Lu sauce in Jiao Liu is also called oil Lu. It is made by pouring hot oil into thickened seasoning sauce. Lu sauce can prolong the penetration of moisture into ingredients. The consistency of Lu sauce should be relatively thick so that it can hold more fat. Only a small amount of hot oil is needed, and it doesn't have to be added into thickened seasoning at a one-go. Lu sauce should be toped on the freshly cooked ingredients immediately after it's prepared to make a sizzling sound and to enhance the infusion of the flavour.

Features of Liu dishes

It is crusty from outside and tender from inside with a strong savoury flavour and rich sauce.

Dishes cooked with this method

Jiao Liu Pork Shreds (Beijing Cuisine)

Sweet and Sour Carp (Henan Cuisine)

Sweet and Sour Mandarin Fish (Jiangsu Cuisine)

Crispy Fish with Lemon Juice (Cantonese Cuisine)

1.3.2 Hua Liu (Moist stir-fried Liu)

Hua Liu, or moist stir-fried Liu is a cooking method that pre-cut and starched ingredients is oil blanched on medium heat and then cooked in Lu sauce. There are Zao Liu (with fermented Zao sauce) and Cu Liu (with vinegar). Zao Liu is to add fermented Zao sauce in seasoning to create a smooth and tender texture and aroma.
Cu Liu is to use vinegar-based seasoning to provide a strong sour taste.

Notes

- Prepare the ingredients with starchy seasoning sauce and leave it rest for a while. Make sure the consistency of starching is appropriate prior to oil blanching so that the ingredients can be separated from each other quickly in the hot oil.
- Starching should be thin and smooth. composition of starching varies. Some ingredients require to dip in egg white while others need cornflour and water mixture.
- Technically, the starching sauce for Hua Liu (Slide Liu) is made in the same way as Hua Chao (Slide Stir-frying), but different in flavour.
- The Ingredients are usually in the shape of shred, slice, strip, dice, or small cubes.

Key skills

- The taste and consistency of the starchy seasoning sauce should be moderate. The Ingredients should be mixed well with the starchy seasoning sauce.
- When oil blanching, the choice of vegetable oil should be made based on the colour and flavour of the ingredients.
- The best oil temperature for oil blanching is between 100°C and 120°C. Too high an oil temperature will cause ingredients to get stuck together while too low an oil temperature will stop starchy coating from firming up quickly around the ingredients. Either way will affect the shape and texture of the dish.
- Most Hua Liu dishes are with a sweet and sour flavour, some are savoury and umami. The marinade should be moderate.

Features of Hua Liu dishes

    Smooth, tender, fresh, aromatic, light but rich in flavour.

Dishes cooked with this method

    Hua Liu Duck Livers (Tianjin Cuisine)
    Liu Vegetarian Fish (Henan Cuisine)
    Hua Liu Fish Slices (Shandong Cuisine)

1.3.3 Ruan Liu (Soft Liu)

Ruan Liu, or soft Liu, is a cooking method in which soft and tender ingredients are cooked by steaming, boiling, or cuan (a type of boiling) first, and then topped with Lu sauce. Soft Liu dishes are mainly fish dishes.

Notes

- Tender and fresh ingredient is the key to a good Soft Liu dish.
- When preparing the ingredients, score cutting and chopping are required where it's appropriate. When steaming, boiling, or cuan-ing, remove the ingredients as soon as they are just cooked.
- In general, oil is not required in making Soft Liu dishes for a rich, aromatic, and umami taste and an extremely smooth and tender texture. On occasions that oil is needed, there should only be a small amount used.

Key skills

- Good score cutting skills and unify-size ingredients.
- Good control over the extent to which the ingredients and starchy sauce are cooked. The ingredients should be removed from wok when it is just cooked. Starchy sauce should be made in accordance with the cooking time of the major ingredient and the balance between the among of starchy sauce used and that of the main ingredient should be appropriate.
- Process ingredients in an appropriate way considering the properties of the ingredients and requirements of the dish. When boiling ingredients, add seasonings such as spring onions, ginger, and cooking wine to remove fishy odour. Always marinate the ingredients prior to steaming remove its fishy odour and to add flavour.

Features of Ruan Liu dishes

    Fresh, tender, smooth, soft, and juicy with a refreshing and rich flavour.

Dishes cooked with this method

    Soft Liu Duck Heart (Huaiyang Cuisine)
    Ruan Liu Bass (Zhejiang Cuisine)
    Ruan Liu Fish Slice (Tianjin Cuisine)

1.4 Bao

Bao, literally translated as pop or explosion, is a cooking method in which firm and crunchy-natured (when cooked) protein ingredients are quickly cooked in extremely hot oil/boiling water or broth on intensive heat. Ingredients are usually pre-cut to bite sized slice, dice, cube. For larger cuts, score cutting is required.

The Bao method is derived from stir frying. The heat-transfer medium is usually oil, sometimes boiling water or broth. Based on the differences in heat-transfer medium, seasonings, and other cooking methods used, Bao can be divided into three types, namely Yóu Bao (Bao with Oil), Jiang Bao (Bao with Paste), and Tang Bao (Bao with Broth).

1.4.1 Yóu Bao (Bao with Oil)

Yóu bao, or Bao with oil, is a type of Bao cooking method that pre-cut ingredients are rapidly cooked by stir-frying in extremely hot wok (sometimes in flame) on intensive heat. Ingredients are often of a tender and firm texture when cooked.

Generally, there are the following two ways to cook Yóu Bao dishes.

- The main protein ingredient is firstly water-blanched, and then oil-blanched in readiness for stir-frying. Other ingredients are then stir-fried in an extreme hot wok on intensive heat, the previously cooked main ingredient is added at this point, followed by adding starchy seasoning sauce, stir all ingredients rapidly using a ladle while repeatedly tossing the wok itself.
- The pre-cut main protein ingredient is oil-blanched prior to stir frying. Other ingredients are stir-fried in extreme hot wok on intensive heat, the previously cooked main ingredient is added at this point, followed by adding starchy seasoning sauce, stir all ingredients rapidly using a ladle while repeatedly toss the wok itself.

Notes

- Bao dishes are best served immediately after cooking. Therefore, the time for serving should be considered when making Bao dishes.
- The consistency of cornflour batter is not recommended to be too thick. Too thick coating would result in pieces of ingredient sticking to each other which will affect the colour and texture of the dish.
- The process of Bao with Oil needs to be complete in rapid sequence without breaks. Intensive heat is required. The amount of oil is usually 2 to 3 times of that ingredients. Not enough oil may affect the flavour of the dish.

Key skills

- Use fresh and crunchy-textured (when cooked) protein ingredients such as squid, cuttlefish, conch, gizzard, and kidney.
- The requirements for cutting techniques applied include cuting the ingredients to a similar size, and the pattern and the depth of each cut should meet the quality standard.
- In terms of heat control, Intensive heat is used in the Bao method. The oil temperature is usually between 180 °C and 210°C. Lower than this range, wouldn't be classified as Bao, higher than this range may cause the ingredients to be brunt from outside yet not thoroughly cooked from inside.
- The starchy seasoning mixture to make the Lu sauce should be prepared prior to cooking. It should be quickly added to ingredients in the wok to thicken. When cooked, the Lu sauce should be tightly coated around the ingredients. A standard to judge if the right tightness of the coating around the ingredients have been achieved is that there should be only just oil left at the bottom of the plate, not the sauce.

Features of Yóu Bao dishes

Tender, crunchy, refreshing, tightly coated sauce and glossy dark red colour.

Dishes cooked with this method

Oil Bao Pork Belly
Oil Bao Dicd Chicken
Oil Bao Squid Rolls

1.4.2 Jiang Bao (Bao with Paste)

Jiang Bao, or Bao with Paste, is a type of Bao that involves rapidly cooking the oil blanched ingredients with pre-cooked paste-based (fermented flour paste, soy bean pastes or fermented bean curb) seasoning sauce on medium to low heat.

Part Three

Notes

- Should select tender, crunchy (when cooked) and fresh protein ingredients as the main ingredient, combined with smooth and crunchy vegetables.
- The consistency of the starchy batter used in Jiang Bao (Bao with Paste) should be relatively thicker than that used in Yóu Bao (Bao with Oil) so that it is easier for the paste-based seasoning sauce to be fully attached to the surface of the ingredients.
- Pay attention to the heat control. The level of heat recommended for Jiang Bao (Bao with Paste) is medium to low. Paste-based seasoning could get burnt on high heat to produce a bitter taste while too low a heat level would not allow the sauce to be attached to the surface of the ingredients. A distinctive feature of Jiang Bao dishes is that only oil but not sauce could be seen at the bottom of the dish when it is finished.

Key skills

- The proportion of paste used should be about 1/5 of the amount of the major ingredient, and the amount of oil needed is about 1/2 of that of the paste. More oil and less paste will not create a rich and even coating around the ingredients; less oil and more paste would easily cause the paste-coated ingredient to burn to the wok.
- The amount of oil is decided accordingly to the consistency of the paste-based seasoning sauce. The principle is that runny sauce requires more oil while thick sauce needs less oil.
- Prior to adding the ingredients in, paste should be thoroughly cooked with seasonings on low heat until.
- Sugar should be added to the dish when it is nearly done. By doing so, it will not only enhance the flavour of the dish but also makes the colour of the dish glossy.

Features of Jiang Bao (Bao with Paste) dishes

Jiang Bao dishes have a glossy dark brown colour, a strong fermented paste aroma, a crunchy texture, and savoury and refreshing taste.

Dishes cooked with this method

Jiang Bao Baby Cuttlefish
Jiang Bao Seafood
Jiang Bao Razor
Jiao Bao Diced Chicken

1.5 Jian (Shallow Frying)

Jian (Shallow Frying) is a widely used oil-based hot dish cooking method wherein flat shaped ingredients are fried in a small amount of oil over medium to low heat until golden on both sides to create a texture of being crispy on the outside and juicy and tender on the inside.

Notes

- Most Jian dishes are single ingredient based. Sometime, fillings are added to the main ingredient.
- When placing the ingredient in wok, use wok shelve to tidy the ingredient up, keep moving the pan and the ingredient in it for even heating.

Key skills

- Fresh, tender and boned protein ingredients are the common choice for Jian (Shallow Frying), Some vegetables are suitable.
- Select appropriate cutting techniques to tenderise the ingredient by hammering to break up its tissue

fibres.
- Ingredients are usually marinated or battered prior to frying.
- Grease the pan, otherwise ingredients will stick to the pan, affecting the colour and the look.
- When frying, move the ingredient around in the wok by tossing the wok. Don't flip over until one side is done.
- Keep a moderate amount of oil in the wok for best frying result.
- Use medium to low heat. The cooking time varies from ingredient to ingredient. Make sure not to undercook or overcook the ingredient.
- Most shallow fried dishes are dry dishes with no juice and can be served directly when cooked.

Features of Jian dishes

Crispy from the outside, tender and juicy from the inside with a golden-brown colour, aromatic, not greasy.

Dishes cooked with this method

Fried Pork Burger
Sichuan Peppercorn and Salt Chicken Burger
Fried Aubergine with Pork Mince Filling
Fried Meat Ball
Jian (Shallow Frying) Cod Steak
Jian Prawn Burger

1.6 Ta

Ta is a combined cooking method in which pre-cut ingredients are seasoned, battered, Jian (Shallow Frying) or Zha (deep fried) until golden on both sides, further cooked in broth with seasonings over medium or low heat. When the food is almost done, thicken the broth by either reducing the liquid on high heat or adding in corn starch. Add a splash of oil before remove from the wok.

Notes
- Dip the ingredient in corn starch and egg right before frying to avoid the coating being affected by emitted liquid.
- Monitor the heat level used carefully to avoid burnt.
- Consider an appropriate plating style to produce a good food presentation. Some dishes may require cutting after cooking.

Key skills
- Ta methods is a combination of Jian (Shallow Frying) or Zha (Deep frying) and one of the variations of braising. Ta dishes should meet the quality standards for Jian and Ta at the same time. For example, ingredients should be pan fried until golden on both sides.
- Ingredients selected to cook with this method should be of easy-cooking, fine and tender properties. This type of ingredients does not require a prolonged cooking time and it is easier to create a crispy and tender texture and a full-bodied taste.
- Dry coating (corn starch or flour) should be even and moderate (not too thick). Egg white coating should be even and thorough.
- The seasoning sauce should be added in at the right time. Thickening with corn starch is an option. It depends on the description of the dish.

Part Three

Features of Ta dishes

Ta dishes are golden in colour, soft, tender, and fresh in texture with a rich and savoury taste.

Dishes cooked with this method

Wok Ta Tofu
Wok Ta Pork Tenderloin,
Wok Ta Yuxiang Pork Slice
Wok Ta Whitebait

1.7 Peng

Peng is a combined cooking method wherein the pre-cut ingredients are marinated, battered or dry coated with corn starch, fried in hot oil on high heat twice with different oil temperature. At this point, the ingredients should be crispy from the outside and tender from the inside. The food would be stir-fried with a seasoning sauce to create the desired flavour.

Based on the initial par-cooking method used, Peng can be further divided into Zha Peng (Deep-frying), Jian (Shallow Frying) Peng, and Chao (Stir-frying) Peng.

Zha Peng deep-fries the ingredient first, and them quickly stir-fry it with a seasoning sauce (non-starchy). Most Peng dishes are cooked with Zha Peng method. Chefs joke that "where there is the Peng, there is the Zha".

Notes

- Larger sized ingredients are usually cut into slices, strips, chunks, or chops. Scoring and hammering applied to protein ingredients such as chicken, beef, pork, and lamb to tenderise them.
- Most ingredients cooked with Peng method should be marinated first. A distinctive feature of Peng is the usage of non-starchy, clear broth based seasoning sauce.
- Peng dishes requires relatively high heat. More specifically, non-battered ingredients should be fried in hot oil on medium heat while battered ingredients should be fried in warm oil on high heat. All ingredients, battered or not, should be removed from oil when they are just cooked, and finally stir fry with a seasoning sauce. At this point, the ingredients are crispy on the outside and tender on the inside.
- Widely used seasonings in Peng method include cooking wine, salt, spring onions, ginger, garlic, spicy soy sauce, fresh lemon juice, sugar, fragrant vinegar. There are a range of combined flavours of Peng dishes include sweet and sour, tomato sauce, savoury and umami, and lychee.
- The larger sized ingredients should be marinated for a longer time to obtain more flavour. Non-battered ingredients do not need marinating. They should be dry coated with corn starch/flour or battered with corn starch solution and/or egg prior to frying.

Key skills

- For non-batter-coated cooking, ingredients should be in the appropriate shapes and uniform sizes. For batter coated cooking, make sure the consistency of the batter is appropriate. If cook with dry coated ingredients, make sure the coat is even.
- The amount of seasoning sauce and the strength of its flavour would have a huge impact on the flavour of dishes. Therefore, it is important to create a balanced flavour. There are a few things to consider, such as if the ingredients are coated, the temperature of the wok, and how easily the ingredients would absorb the flavour of the seasoning sauce.
- Broth based seasoning sauce should be prepared prior to the frying. This will not only shorten the cooking time, but also allow seasonings in the sauce to release their flavours to make the combined flavour nicer and stronger.

- Oil temperature should be carefully controlled when frying. In Peng method, ingredients are usually twice fried. The first time is for crusting while the second time is for cooking. Relatively high oil temperature is required in Peng.

Features of Peng dishes

Peng dishes are aromatic from the outside and tender from the inside and slightly juicy with a refreshing taste. It is an ideal dish for summer.

Dishes cooked with this method

Peng Prawn

1.8 Shāo

Shāo, or braising, is a cooking method wherein ingredients are par-cooked by using another methods such as deep frying, shallow frying, stir-frying, or blanching, and then cooked in broth or water with seasonings initially at high heat to bring it to a boil, then reduce to medium to low heat until broth/water is thickened.

A wide range of ingredients are suitable for the Shāo (Braising) method. Shāo dishes present a variety of flavours. The size of the cooked item varies. There are also different requirements for the extent to which ingredients should be cooked. For example, Shāo fish is required to be just cooked but for Shāo Pork/Beef/Lamb, it should be well cooked until the ingredient becomes tender and soft.

According to the colour of the cooked dish and the techniques involved in cooking, Shāo (Braising) can be divided into Hong Shāo (Braising in Red Sauce), White Shāo (Braising in White Sauce) and Dry Shāo (Dry Braising).

1.8.1 Hóng Shāo (Braising in Red Sauce or Red Braising)

Hong shao, literally means red braising, is a cooking method wherein the pre-cut ingredients are par-cooked by using another method such as blanching, deep frying, shallow frying, stir frying, or steaming. The par-cooked ingredient is then cooked in broth or water, firstly on high heat to bring it to a boil, followed by removing scum, adding seasonings in including caramel and soy sauce, reducing to medium or low heat, cooking until the ingredients are soft and tender, and liquid is thickened. Sometimes adding corn starch to further thicken the liquid is required. Hong Shao dishes are umami, tender, fatty but not greasy. The techniques involved are shown in the flowchart below:

Selecting ingredients → cutting → par-cooking → braising with seasoning including caramel and soy sauce to create the red colour → thickening by adding in corn starch solution (occasionally by reducing liquid on heat only) → plating

Notes

- Ingredients should be processed to remove the odour, soften the texture for a better marinade-infusion.
- In the par-cooking process, the following should be taken care of:
    - Avoid moisture loss when frying or stir-frying.
    - Browning needs to be done evenly.
    - Control the level of heat used carefully to avoid burning.
- Seasoning must be done in the right sequence and at the right time. When red braising protein ingredients, cooking wine should be added first to remove the odour, then follow on with soy sauce, salt, sugar and other condiments and broth. When adding broth or water in, the appropriate amount should be poured in along the walls of the wok. Water should be used instead of broth when red braising the aquatic ingredients to maintain the natural umami taste. For vegetable ingredients and rehydrated dried-ingredients, umami soup (such as chicken soup) is often used. For protein ingredients such as meat and poultry, it's best to use broth made of the same ingredient. For example, pork

broth is the best for braising pork.
- Skilfully control the extent to which the ingredient is cooked and add in cornflour liquid at appropriate time.
  - Decide on the amount of cornflour liquid to be used based on the amount of liquid left in the wok. Cornflour liquid needs to be well-mixed without clots. When adding corn starch solution into large cuts or whole sized ingredients, keep tossing the wok. cornflour liquid should be poured slowly onto the broth, not onto the ingredients.
  - Adding a splash of oil is the final step. Again, the oil should be poured in along the walls of the wok, keep tossing the wok to make sure the oil and broth are mixed well. This step is to give the dish a glossy colour.

Key skills
- To make the red braised dish look tidy, small sized seasonings such as ginger, spring onion, and Sichuan Peppercorns and herbs should be bagged. Small bits of broad bean paste, after stir-fried, should be removed as well.
- To ensure the quality of braised dishes, par-cooking shouldn't be done too early prior to braising. It may affect the colour, aroma, taste and presentation of the cooked dishes.
- During the braising process, make sure ingredients are not stuck to the wok. Chicken bones or pork bones can be laid at the bottom of the wok to avoid it.
- If ingredients are of different texture or types, adjust the extent to which each type of the ingredients are cooked when par-cooking, or add them in sequence when braising.
- For tender ingredients, the required cooking time is short. The quality expected include that the ingredients are just cooked with a smooth and tender texture, moderate amount of broth left in the dish, and fully developed flavour.
- When braise tougher ingredients that require a longer cooking time, carefully consider the texture of the ingredient to decide on the amount of broth, the length of the cooking time, and the level of heat. Must avoid trying to shorten the cooking time by braising on greater level of heat with larger amount of broth.
- Select appropriate seasonings, such as soy sauce, broad bean paste, cooking wine, red wine, flour paste, tomato sauce, sugar, to add colours to the ingredients. The choice of colour should be in according to with the intensity of the taste. As a result, each combined flavour will have a colour shade to match with. For example, sweet and salty flavour goes with a red orange colour, savoury and umami flavour pairs with a soft yellow, home-made flavour matches with golden red and five-fragrance flavour comes with golden yellow.
- Reducing the broth is the key to produce a richer taste. It also can enhance the colour of a dish and the brightness of it. Before reducing, make sure there is enough broth left in the wok to avoid the food being burnt.

Features of Hóng Shāo dishes

Red braised dishes are soft, tender with thick broth and a rich taste. The colour is usually glossy dark brown.

Dishes cooked with this method

Red Braised Pork
Red Braised Fish
Red Braised OX Tails
Braised Fish in Red Sauce

1.8.2 Gan Shao (Dry Braising)

Compared to Red/White Braising, Gan Shao (Dry braising) only uses medium or low heat to reduce the broth. The use of corn starch to thicken the broth is next to none. No broth should be seen after cooking is finished. As a result, the flavour is fully captured by the ingredients instead of infused in the broth.

Although Gan Shao is generally regarded as a signature cooking method of Sichuan Cuisine, it is widely used in other regional cuisine styles such as Shangdong cuisine. To create a typical Sichuan flavour, the most common used seasonings for gan shao dishes are cooking wine, pickled red chili pepper, Pixian broad bean paste, spring onions, ginger, garlic, sugar, fermented rice, and fragrant vinegar.

The techniques involved are shown in the flowchart below:

Selecting ingredients → initial processing → cutting → par-cooking → braising with seasoning → thickening by reducing (the use of thicken agent is rare) → plating

Notes

- Broth should not be seen in the dish after cooking is finished. The dish should have a rich flavour and it looks moist and glossy, not dry and dull.
- Control the heat level properly.
- High heat is used for stir-frying the ingredients and bringing the broth to a boil; medium and low heat should be applied afterward to allow flavour infusion in the broth; high heat should not be used for ingredients that are gelatine rich or too soft to be flipped over. To keep food from sticking to the wok, constantly tossing the wok while braising.
- Add seasonings in the right sequence and in the right proportion. For some special seasonings (such as flour paste) should be stir fried on medium heat until aromatic, and then mixed well with broth prior to adding in the ingredients.
- All seasonings should be finely chopped. Spicy broad bean paste should be stir-fried in warm oil on medium heat until red oil emitted, add the broth in, bring it to the boil, remove the scum, and then plunge in the ingredients.
- The use of vinegar in dry braising needs to be moderate. The purpose of using vinegar is to dismiss the bad smell and enhance the aroma, not to give the sour taste. Normally vinegar is added in right before the cooking is finished.
- Gelatine-rich, tender, and fresh ingredients are ideal for dry braising. If use dehydrated ingredients, make sure the extent to which the rehydration is properly done, ideally, they should be soft but firm. Prawns, fish, chicken and vegetables should be well-washed to minimize the bad odours and remove elements that may affect the texture of the dish.
- Ingredients are either cut into strips and chunks or cooked in their original shape, prawns, fish, chicken and vegetables should be oil-blanched prior to braising to help maintain the shape, increase the fragrance and make them easier for cooking.
- The popular flavour combination of dry braised dishes includes savoury-umami, home-made flavour, and aromatic paste flavour. Dry braising should always use medium heat to bring the broth to a boil, and then continue cooking on medium to low heat prior to increase to medium heat again to thicken the broth.

Key skills

- Par-cooking affects the colour, aroma, taste, and shape of the dishes. Therefore, make sure the right processing method is used. For example, marinated fish and prawns should be blanched in hot oil to crust them so that their taste and shape won't change a lot when braising. Vegetables should be blanched in warm oil to keep the colour and freshness. Ingredients such as tendon and sea cucumber are hard for flavour exchange but wouldn't stand for prolonged braising time hence, they should be

stewed before braising.
- The amount of broth used should be moderate. It should be decided based on the property of the ingredients and the time required for braising. Generally the tender, moist and easy-to-cook ingredients need less broth and vice versa.
- Seasonings used in dry braising are plenty including cooking wine, caramel, vegetable oil and broad bean paste. They are the crucial elements for the intense and aromatic flavour of braised dishes. To ensure the best effect of these seasonings on the colour, aroma, and taste of the dishes, chefs must understand the colour shades these condiments possess, the flavour combinations they can create, and the right sequence of adding them in.
- Steep ingredients in broth so that they can fully absorb the flavour. When reducing the broth, keep moving the ingredients for an even flavour infusion. For fragile ingredients (like fish), do not move them, instead scoop the broth and pour on top of the ingredients to help an even flavour infusion.

Features of Gan Shao (Dry Braising) dishes

Gan Shao dishes are golden brown in colour with a moist and glossy appearance, a soft and tender texture, not greasy. Every piece of the food should be coated evenly with the thick broth. The broth is free from starch. The taste can be best described as aromatic, umami, clean, and rich.

Dishes cooked with this method

Dry Braised Beef Briskets
Dry Braised Carps
Dry Braised Winter Bamboo Shoots
Dry Braised Butterfishs
Dry Braised Abalonees
Dry Braised Crucians

1.9 Mènn

Mènn, a variation of braising, is a cooking method that ingredients are par-cooked by using another method such as deep frying, pan frying, stir-frying, or blanching, and then seasoned with soy sauce, sugar, spring onions, and ginger. The next step involves adding in broth and bringing it to a boil on high heat, removing scum, adding in further seasonings, and simmering on medium to low heat with the lid on until cooked. Finally, high heat is used again to thicken the broth.

Two key aspects of culinary techniques that are highlighted in Mènn, the control of heat level used and blending of flavour.

The cooking time is adjusted based on the size and texture of the ingredients. The differences between Mènn and Wei are subtle and similar to that between braising and stewing. In general, Mènn dishes have less broth and require less cooking time.

Mènn has four variations: Huang Mènn (Mènn with Yellow Sauce), Hong Mènn (Mènn with Red Sauce), Jiu Mènn (Mènn with Alcohol) and You Mènn (Mènn with Oil). Although the seasoning, ingredients, and par-cooking techniques involved in each may vary, these four variations share the same cooking process.

Techniques of Mènn method include the following:

Selecting ingredient → cutting → par-cooking → seasoning and Mènn → thickening → plating

Notes

- Suitable ingredients for Mènn method are collagen-rich and easily cooked with a firm and tougher texture. Protein ingredients such as beef, pork, shanks, chicken, duck, soft-shelled turtle, and see eel are commonly used. These ingredients can be cooked in its original form or cut into strips, dices, and

chunks. Root vegetables such as winter bamboo shoot, Chinese leaves, and celtuce are popular as well.
- In term of the heat control, three levels of heat are used:
  - High heat is used at the beginning for browning and removing bad odours of ingredients.
  - Low heat is then applied to simmer ingredients over a long period of time until soft and tender.
  - At the end of the cooking, high heat is used again to thicken the broth. The colour of the dish goes richer as a result.
- For seasoning, the initial seasoning involved in the par-cooking should use spices and herbs that help to remove odour and enhance the flavour and aroma. Further seasonings are added at a later stage to define the colour and flavour of the dish.
- The maturity of the ingredient and the level of collagen contained in it, as well as the required texture of the dish are the factors to consider when deciding on if starch is needed. Before serving, make sure the oil, starchy seasoning sauce, and ingredients are well proportioned. Thickening can be achieved by reducing the liquid or adding in starch. Whichever way is chosen, the amount of oil used needs to be monitored carefully. Too much oil would separate the seasoning sauce from the ingredient. Too little oil may cause the ingredient being dry, burnt, or unevenly coloured.

Key skills
- The Choice of par-cooking method is based on the texture of ingredients.
- Use enough broth to submerge the ingredients completely. For tougher cuts, a bigger volume of broth should be considered, and vice versa. To simmer easily cooked ingredients, medium heat should be used. For tougher cuts, it is best to be cooked over a longer time on lower heat to get the soft and tender texture. It is not recommended to frequently lift the lid to check the ingredients as it will let heat to escape which will affect the colour, aroma, and taste of the dish. Meat and poultry Mènn dishes are best to be garnished with a green leafy vegetable bed. Not only it will enhance the aroma but also reduce the greasiness.
- In the initial seasoning, make sure browning is done evenly. When use seasonings to define the flavour of the dish, Mènn with yellow sauce and Mènn with alcohol dishes should have rich, savoury and umami tastes with strong aroma. Mènn (post roasting) with Red Sauce dishes are dominated by home-made flavour which is rich and spicy. Mènn with oil is glossy red in colour with a complex flavour of refreshing, savoury and umami.
- The cooking time and amount of broth used should be calculated carefully. The heat intensity applied should be determined by the required texture of the dishes. Adding broth in the middle of simmering is not encouraged.

Features of Mènn dishes

Once cooked, the shape of the food is remained. The dish is soft, tender, aromatic and savoury with rich broth.

Dishes cooked with this method

Mènn Chicken with alcohol
Mènn Lamb Rib with Yellow Sauce
Mènn Fish with Yellow Sauce
Mènn Lamb with Red Sauce
Mènn Pork and Pumpkin with Alcohol

Part Three

1.10 Kao

Kao is a cooking method that ingredients are par-cooked by deep frying, pan frying, stir-frying, or blanching, and then seasoned with broth and spices, simmered in a covered wok for an extended period of time on low heat to allow the broth to thickened by reducing. The ingredients would have fully absorbed the flavour of the broth. Both cooked and raw ingredients can be used in this method.

Techniques involved include the following:

Selecting ingredients → cutting→ par-cooking → seasoning and Kao cooking → thickening the broth on low heat → plating

Notes

- Ingredients should be fresh and easy-to-cook type such as chicken, duck, fish, and prawns. They are usually cut into chunks, thick slices, or strips of the same sizes for even heating.
- Ingredients should be par-cooked by frying, steaming, or boiling. The Kao process allows ingredients to absorb the flavour of the broth. The Right amount of broth to start the Kao process is the key.

Key skills

- The Kao process is for flavour exchange between the ingredient and the broth.
- The Kao dishes should be garnished with green vegetables when serving.
- When the Kao process is finished, pick spices such as green onion and ginger to make the dish looks neat and clean.

Features of Kao dishes

Rich broth, savoury taste, glossy colour.

Dishes cooked with this method

Kao Duck

Kao King Prawns

1.11 Dùn

Dùn involves placing the prepared ingredients in water or broth in a pot or other earthen utensil, bringing it to a boil on high heat, and then simmering on low heat until ingredients are soft and tender.

Dùn dishes are featured in umami broth, natural tastes of the ingredients, intact shape, and a tender but not fragile texture. Unseasoned Dùn with clear broth is called clear Dùn and the thick broth type is called mixed Dùn. They share the same cooking process but produce different tastes.

1.11.1 Double Dùn

Double Dùn is a cooking process wherein the prepared ingredient sits in a container and placed on top of a base inside a bigger pot (outer) with water to steam, like a double jacket boiler.

Notes

- Double Dùn takes a long cooking time and is inefficient in eliminating unpleasant odours. Therefore, fresh, fibre-rich and tendon-rich, and tough cuts ingredient are an ideal choice for it. Prior to cooking, rinse the ingredients well and cut them into large chunks, or keep them intact.
- Due to the use of indirect heat, cooking by double Dùn is long and slow. The utensil should be covered to seal the fragrance, umami and moisture and by doing so the natural taste and shape of ingredients remain.
- Ingredients should be blanched to remove blood and any scum and rinsed prior to placing in the inner clay pot or other earthen utensils. Sufficient water or broth and seasonings should be added, and

the covered pot then should be fitted into a bigger outer pot to steam.

Key skills
- Ingredients should be blanched to remove the fishy, gamy or other odours and impurities.
- Ingredients should be rinsed and then put in an inner clay pot with broth and seasonings that help to remove unpleasant odours. The pot is then placed in the water in an outer pot. Water in the outer pot should be lower than the height of the inner pot to avoid water getting into the inner pot when boiling.
- Clay and ceramic utensils are preferred because they are good at heat transfer.
- Seasonings should be added after the ingredients are fully cooked, soft and tender. Once seasoned, serve immediately.

Features of Double Dùn dishes

Clear broth, natural and umami tastes, tender texture with an intact shape, aromatic.

Dishes cooked with this method

Old Duck Dùn Caterpillar Fungus*
Steam Pot Chicken
Turtle Dùn Chinese Yam

* Caterpillar Fungus (literally "summer grass, winter worm"), or dōng chóng xià cǎo (Chinese: 冬虫夏草)

1.11.2 Direct Dùn

Direct Dùn, as the name implies, is a cooking method where the prepared ingredients and seasonings are placed in an earthen pot with water which is then brought to a boil on high heat and simmered on medium to low heat until ingredients become soft and tender.

Notes
- Use fresh, tough protein ingredients or large root vegetables for cooking.
- Blanch and rinse the ingredients prior to Dùn.
- Add enough broth at the beginning of Dùn and cover the pot with it lid. Not to open the pot frequently during Dùn.

Key skills
- Control the levels of heat properly: use high heat at the beginning, and then reduceit to medium or medium to low heat for simmering, and finally further reduce to low heat in the end.
- Prevent the earthen pot from cracking by using moderate amount of water.
- Serve immediately after cooking.
- Season before serving.
- No thickening agent is required.

Features of Direct Dun dishes

Clear soup, natural tastes, soft and tender with strong aroma.

Dishes cooked with this method

Dùn Chicken with Mushrooms
Mixed Dùn
Home-Made Dùn Chicken

1.12 Zheng

Zheng, or steaming is a cooking method wherein pre-cut and marinated are cooked with steam as the temperature in the steamer reaches its peak, and generates certain pressure. It can heat the ingredients evenly and keep their moisture. Steaming is widely used, be the ingredients large or small, whole or cut, tender or tougher. The technique of heat control in steaming is about choosing the right intensity of steam, the steaming time, and type of steamer, all of which depend on the properties of ingredients and quality requirements of the dish. Below is a summary of heat control considerations.

Table 10 Heat Control Consideration

Steaming methods	Properties of the Ingredient	Quality requirements of the Dish	Intensity of heat	Steaming time	Sample Dishes
1	Fresh and Tender	Just cooked	High heat on boiling water	Instant	Clean Steamed Fish Crush-rice Coated Steaming Beef Tenderloin Slices
2	Tougher large cuts	Soft and tender	High heat on boiling water	Extended	Official Court
3	Tender and delicate	Maintain the shape of the ingredients	Medium to low heat	Extended	Orchid-shaped Pigeon Eggs Shao Mai with Egg Wrapper Steamed Duck Feet

Whatever steaming approach is adopted, the amount of level added into the steamer is a key. If insufficient, the ingredients may become yellow and wither, and hard to cook. When using a multi-layered steamer, there are few rules of thumb to follow.

- Ingredients may emit liquid that should be placed in the bottom steamer while the dry ones in the top steamer.
- Light-coloured ingredients should be in the top steamer while dark ones at the bottom steamer to avoid light ones being stained.
- Easily-cooked ingredients should be in the bottom steamer and those hard-cooked be in the top steamer.
- If the quality requirement is to maintain the natural flavour of the ingredient, leave it in a covered pot, and then steam the pot to avoid direct contact with liquefied steam which will affect the colour and aroma of the ingredients.

Steaming can be divided into Qing Zheng (Plain Steaming) and Fen Zheng (Crushed Rice Coated Steaming).

1.12.1 Qing Zheng (Plain Steaming)

Qing Zheng, or Plain Steaming can be done in two ways. Place par-cooked ingredients with seasonings in a flavoured broth, in a pot, and then steam the pot. This way is also known as double steaming; or to steam seasoned ingredients directly.

Notes

- For Plain Steaming, ingredients should be very fresh without any undesirable odour. The hair, if any, should be picked or singed and bloodstains should be removed.
- Meat and poultry ingredients usually remain intact or cut into large chunks. They take a long time to be steam cooked, and should be blanched, rinsed first.
- Fish ingredients are to be chopped, or remain as a whole with crosswise scores, and steamed at intensive heat for a short time.

- For double steaming, put the ingredients in a pot with seasonings and umami-flavoured broth in it. A common flavour combination is savoury and umami.
- The heat control technique is a key to steaming. Please refer to table 1 for more detail.

Key skills

- To steam a few dishes using one multi-layered steamer, the clean-steamed dishes should be placed in the top steamer to avoid colour stain and flavour exchange with other dishes.
- Remove the ginger, spring onions and Sichuan peppercorn prior to serving to keep the dish cooking tidy.
- The extent to which ingredients are cooked in the blanching process should be carefully monitored. Rinse well the ingredients after being removed from the boiling water. Cut them when they are cooled down for easy shaping.
- Choose the right steaming method according to the required texture of the dish. Dishes steam-cooked on intensive heat for a long time can be kept in a warmer before serving. Dishes are steam-cooked instantly should be served immediately.

Features of Plain-steamed dishes

Plain-steamed dishes brought out the natural flavour of the ingredients, which was very refreshing. The texture is either soft and delicate or soft and tender with considerable amount of liquid.

Dishes cooked with this method

Plain-steamed Shad
Plain-steamed Whole Chicken
Plain steamed Mandarin Fish

1.12.2 Fen Zheng (Crushed Rice Coated Steaming)

Feng Zheng, or Crushed Rice Coated Steaming, is a cooking method where marinated ingredient ingredients are coated with crushed rice and then steam-cooked until soft and tender.

Key techniques involved include the following:

Selecting of ingredients → cutting → marinating → mixing with crushed rice → shaping/Wrapping → steaming

Notes

- Ingredients are suitable for Crushed Rice Coated Steaming are tougher cuts without tendons, fresh and flavourful with both fat and lean; or tender without tendons, umami, delicate and easy to cook. Ingredients are best cut into strips, slices and chunks.
- Ingredients should be marinated first to absorb the flavour of seasonings. A few popular flavours combinations include savoury and umami, salty and sweet, five fragrances. The features of a dish should be considered when determining the flavour combination.
- The proportion of crushed rice to ingredients may differ according to the texture and fattiness of the ingredients. In general, it should be somewhere between 0.06 : 1 and 0.1 : 1.
- Some ingredients are wrapped in lotus leaves while others are steamed directly. Ingredients should be stacked loosely together to allow even heating.

Key skills

- The marinating time should be appropriate. It should be in agreement with the quality requirements of the dish, the texture and size of the ingredients.
- To make crushed rice, stir fry rice on low heat until light golden.
- Grind after the rice cools down. Make sure the steps are followed in the right order to achieve the best pulpy, gummy and grainy texture of the rice coating.

- For less fatty ingredients, add some oil during seasoning so they will become moist and gummy.
- After coating with crushed rice, the ingredient should be moist but without visible liquid in it. The required texture and taste of the dish can be described as soft and tender, aromatic and umami, flavourful, gummy but not greasy.
- Crushed Rice Coated Steamed dishes should be cooked without interval. Avoid heat reduction or heat loss during the process as it will cause liquified steam getting into the dish to compromise the quality of it.

Features of Crushed Rice Coated Steamed dishes

They are coloured in light golden or glossy golden with a gummy and moist texture

Dishes cooked with this method

Crushed Rice Coated Steamed Beef
Crushed Rice Coated Steamed Lamb
Crushed Rice Coated Steamed Pork Belly in Lotus Leave

1.13 Hui

Hui, is a cooking method wherein pre-cut ingredients are the shape of slices, shreds, strips, dices, and cubes is cooked in a wok with seasonings and broth on medium to high heat for a short time and then thinly thickened with corn starch solution.

Hui includes Qing Hui (Plain Hui) and Bai Hui (White Hui).

The key techniques involved include the following:

Selecting ingredient → preparing ingredient → cutting into shapes → par-cooking → stir-frying until aromatic → hui-ing → plating

Notes

- Ingredients suitable for the Hui method are best to be previously cooked, par-cooked or easy to cook type. For example, rehydrated dry ingredients, par-cooked pork balls, prawn balls, and fish balls. Hui dishes usually contain more than two ingredients.
- The Hui method requires a short heating time hence the ingredients should be cut into smaller size for easy cooking.
- Fresh broth is the main seasoning for hui dishes. When deciding on the type of broth to use, the properties of ingredients and the quality requirements of the dish should be considered.
- Hui dishes are often thinly thickened on intensive heat so that corn starch can absorb water to the full.

Key skills

- Ingredients pairing should be carefully made based on the colour, aroma, taste, texture of the ingredients and the proportions of proteins to vegetable to highlight the uniqueness of Hui dishes.
- For ingredients with bland taste or odour, for example rehydrated dry sea cucumber and squid, stew them in fresh broth first; for heat dedicate ingredients such as tomatoes, broad beans, and choy sum, add them towards the end of the cooking process. Par-cook the ingredients with water blanching or oil blanching to the level of just cooked.
- The Hui-ing time should be appropriate. When the flavour combination is determined, quickly thicken dish by adding in corn starch solution, and remove from the wok immediately to shorten the Hui-ing time to best preserve the aroma and umami taste. There is no fixed standard on consistency of the corn starch solution. However, it shouldn't make the dish look soggy and the colour and flavour of the dishes are not overshadowed.

- Different textures of the ingredients determine that they should be added in the right order. The consistency of corn starch solution should be too thick and must be quickly stirred in the wok to avoid clotting.
- The process of making Hui dishes includes stir-frying spring onion and ginger until aromatic, add fresh broth and bring it to a boil, remove spring onions and ginger, add the ingredients and seasonings, bring it to boil again, thicken with corn starch solution, and serve.

Features of Hui dishes

A wide range of ingredients are suitable for the Hui method. Hui dishes are made with fresh broth and served in the resultant gravy. They have rich colours, a refreshing and umami taste, and asmooth texture. Some hui dishes would be considered as soup dishes.

Dishes cooked with this method

Lamb Hui Noodles
Song Sao Fish Soup (Aunty Song's Fish Soup)

1.14 Cuan

Cuan, Zhu and Shuan are water-based cooking category. Influenced by today's heathy eating concept, people tend to go for a lower oil consumption option if they have the choice hence these three water-based cooking methods have gained more popularity. Control of heat is believed to be the key element affecting the quality of dishes cooked with Cuan, Zhu, and Shuan.

Cuan, or quick boiling, requires to coat the pre-cut ingredient with corn starch solution (occasionally this step is not needed), and then quickly cook them in boiling broth or water on high heat. Alternatively, ingredient can be processed into balls which will be par-cooked first by using another method, and then quickly cooked in boiling broth or boiling water on high heat to finish off.

The process of Cuan technique is shown in the flowchart below:

Selecting ingredient → cutting or mincing → mixing with corn starch solution or making balls → Cuan-ing (cooked quickly in boiling broth or boiling water on high heat) → serving

Notes

- Selecting ingredients: ingredients should be fresh without unpleasant odour, tender and delicate.
- Cutting or mincing the ingredient: remove tendon if any, cut the main ingredient into asmall size in the shape of shred, slice, dice. Alternatively, mince or blend the main ingredient until smoot. Other ingredients, for example vegetables, should be selected for easy cooking e.g. leafy types to be processed into appropriate shapes.
- If the ingredient is coated with corn starch solution, the consistency of corn starch solution should be in line with the quality requirements of the dish.
- For example, pork emits liquid when being heated hence it gets tougher and dry. To achieve a tender and smooth texture, pork should be mixed with conflour and water prior to cooking. It not only compensates the water loss during the cooking but also reduces the amount of liquid being lost by providing a coating.
- Prawn meat, on the contrary, is naturally tender and contains more liquid. It requires less cooking time as well. Adding in a small amount of dry corn starch will be enough to achieve a similar tender and smooth texture.
- Cuan-ing: there are two approaches of Cuan. One is called clear broth Cuan that the main ingredients are mixed with corn starch solution prior to plunging into boiling water remove when colour is changed, plunge back in to boiling broth and serve. The other one is called mix-in-the broth Cuan in

which ingredients are cooked in boiling water with seasonings, remove scum if any. Once cooked, the ingredients are placed in a serving bowl with freshly made broth and served.

Key skills

- Cuan dishes are highly demanding in terms of cutting technique. Ingredient should be cut in to pieces of similar size and thickness. Otherwise the appearance of the dish and the length of the cooking time may be affected. Ingredients should be fresh, tender and delicate with less odour, any bone, skin and tendon should be removed.
- Ingredients should be blanched first. For an easy and even cooking, ingredients are best to be sliced, shredded, strips or in a small ball shape of similar size and thickness.
- Ingredients a tender and moist nature are to be coated with corn starch solution prior to Cuan to maintain the texture.
- Seasonings are to be added either prior to or after the blanching. Thickening by corn starch is not needed.
- To mince protein ingredients, trim off the remaining connective tissue and chop the meat finely.
- To make meatballs, add corn starch to minced meat and mix thoroughly until the meat holds together firmly. To Cuan the meatballs, bring water to the boil first. Cuan in not-boiling water will result in uncoated ingredients getting tough and coated ingredients to lose their corn starch coating. This will hugely affect the quality of the dish.
- If minor ingredients are selected, make sure the proportion does not overshadow the main ingredient to keep the tender texture of the dish. For those further ingredients that take a longer to cook, blanching should be used prior to Cuan to shorten the cooking time of the dish.
- To Cuan non-coated ingredients, remove scum if any to maintain a clean and nice look of the dish. To Cuan corn starch coated ingredients, plunge in the boiling broth/water, stir to separate the pieces from each other, remove when the colour of the ingredient changes. Place in a bowl, add in seasoned fresh clear broth with a smooth texture.

Features of Cuan dishes

With clear soup, Cuan dishes usually have a clean and umami taste and a tender and smooth texture.

Dishes cooked with this method

Clear-soup Cuan Meatballs
Milky-soup Crucian
Cuan Fish Slice with Tofu

1.15 Zhu

Zhu, or boiling, is a cooking method in which pre-cut and/or par-cooked ingredient is plunged in to broth, bring it to a boil on high heat first, and then cooked over an extended time on medium to low heat to the ingredient thoroughly over an extended time.

In Sichuan Cuisine, there is a method called water-boiling, in which pre-cut protein ingredient, such as chicken, fish, pork, or beef are marinated and mixed with corn starch solution, and then oil blanched prior to boiling in seasoned broth until cooked, thicken the broth by reduction (some ingredients may require thickening by adding corn starch). Meanwhile, stir fry fresh vegetables to lay at the bottom of a serving bowl, pour the main ingredient with broth on top, sprinkle chopped chili pepper and Sichuan peppercorn powder, finish by adding in hot-tempered oil. Mala dishes ("Mala", literally translated as "numbing and spicy" is a popular flavour combination in Sichuan Cuisine) such as Water-boiled Beef, Water-boiled Pork, Water-boiled Fish, and Water-boiled Chicken Breast are made in this way.

The Process of Boiling technique is shown in below.
Selecting ingredient → preparing or/and cutting ingredient → boiling → seasoning → serving

Notes
- In selecting ingredients: the defining flavour of the boiled dishes dominated by the natural taste of the ingredients. It is so because the substances of the ingredients that impart umami will dissolve in broth when heated over a period of time. For this reason, protein ingredients should be fresh, less gamey or fishy, and rich in protein and fat. Vegetarian ingredient should be heat-resist and protein-rich.
- In preparing the ingredients: with game and fish, water-blanching and oil-blanching should be used to minimise the odour.
- In boiling: In order to have rich and white broth, Boiled Dishes should be cooked in high heat. As a result of clear and umami broth, low heat should be used for cooking. Small sized and tender ingredients should be cooked instantly on high heat; equally, large sized and tough ingredients need to be cooked over a longer period of time on low heat. Sufficient liquid should be added at the beginning of the cooking. Adding extra water or broth during cooking should be avoided. When using this method to cook fresh aquatic ingredients, water is preferred over broth; while making meat, poultry, or bean dishes, chicken broth or thick and white bone broth should be the choice. If necessary, add some lard into the broth to for a whiter broth. Keep the lid on when boiling.
- Seasonings: Seasonings should be added towards the end of the cooking. Normally it is the last step prior to removing from the wok. If seasonings being added too early during the cooking, it takes longer time for ingredients to be cooked. In some cases, the protein is prevented from being dissolved in liquid easily.

Key skills
- In preparing the ingredients: use aromatic spices such as ginger, spring onions, and Sichuan peppercorn to remove odour and add aroma. These spices should be removed from the dish before serving to provide a neat and clean presentation.
- In boiling: ingredients are best to be cooked in the shortest time possible to maintain good colour, aroma, and the taste of the dish. Therefore, ingredients should be soft and tender, and cut into a similar size and thickness.
- Boiled dishes should be removed from heat once it's done to avoid being overcooked.
- Liquid and ingredients should be well-proportioned.
- In selecting ingredients: ingredients should be fresh and easy to cook. Tough skin of vegetables should be removed. Protein ingredients such as pork and poultry should be cleaned (free from hair and blood stain) and cut neatly into shreds and slices. Some vegetables will require par-cooking prior to boiling. Fish ingredients should be in chops and chunks or remain intact.

Features of Zhu (Boiling) dishes

Boiled dishes come with a large amount rich broth soup so sometimes are served as a soup dish. The taste is clean and umami.

Dishes cooked with this method

Tai-an Boiled Tofu
Water-boiled King Prawns
Water-boiled Pork Slice
Water Boiled Beef Slice

1.16 Shuan

Shuan, or Chinese fondue, is a DIY cooking and eating method. It involves bringing the seasoned base soup to a boil in the pot on the dining table, plunge pre-cut ingredients in the boiling or simmering soup until cooked, consume with dipping sauce. The options of base soup and sauce are many. For example, clear soup, white soup(bone broth), umami soup, spicy soup etc. Dipping sauce can be customised to individual taste. The essence of Shuan method is to allow dinners to make their own dipping sauce and control the cooking time.

Shuan technique requires the skills to fulfill the following steps:

Selecting of ingredients → processing fresh ingredients → slicing frozen ingredients→ making dipping sauce → Shuan

Notes

- Ingredients should be cut into the size and shape that are convenient in Shuan, for example, large and thin slices. Small sized ingredients such as prawns and clams are best to be skewered.
- Base soup/broth in the hot pot should be kept boiling. Regularly add in more soup/broth to maintain the amount of liquid in the hotpot. Basic clear base soup is seasoned with cooking wine, salt, ginger and green onion infused water.
- Options of seasonings to make the dipping sauce is a mini-buffet on its own. Commonly provided seasonings include sesame paste, peanut paste, fermented bean curd, soy sauce, spicy soy sauce, chili oil, shrimp sauce, fragrant vinegar, salt, white pepper powder, spring onions, ginger, and garlic. Other options include fermented leek flowers, coriander mince, sugar-pickled garlic etc.

Key skills

- When selecting lamb, the texture, odour, and freshness of the cut should be considered. Connective tissues, tendons, and soft bones should be removed. Lamb should be thinly sliced against the grain.
- Keep adding additional soup in to the hot pot to maintain the level of liquid and the flavour of the base soup.

Features of Shuan dishes

A wide range of ingredients are suitable for Shuan. There are many flavour combinations. The texture of cooked food is open soft and tender with umami soup. The unique dining style of eating while cooking is a huge bonus to the dining experience.

Dishes cooked with this method

Beijing Style Shuan Lamb
Chongqing Style Hot Pot
Guangzhou Hot Pot (Cantonese style)

1.17 Basi (Caramelising/Candied)

Basi dishes are candied dishes. Basi technique literally means pulling into threads. It involves deep frying food in the shape of chunks, balls or strips, making sugar syrup, mixing the food with the liquid sugar and swiftly pulled food pieces apart to create brittle threads between them. Heat control skill is a key to this technique.

Notes

Depending on the heat medium used to create sugar syrup, Basi can be divided into Shuǐ Ba (Water Caramelising), Yóu Ba (Oil Caramelising), and Shuǐ-Yóu ba (Water and Oil Mixed caramelising):

- Shuǐ Ba (Water Caramelising): Sugar is heated on medium to low heat with water until it turns brown. In which case, water is the heat transfer medium. As the boiling point of water is 100°C, the caramel

would not get dark easily. Therefore, water caramelising tends to create a light brown crystal coating.
- Yóu Ba (Oil Caramelising): Oil is used as the heat medium to melt sugar. Oil caramelising's characteristics include instant heat transfer, hot in temperature, and oil temperature increases quickly. The caramel was made with this technique is bright and glossy, which creates fine and long sugar threads. However, this is a relatively difficult technique as the oil temperature increases quickly and its boiling point is high, sugar is be easily gets burnt. The amount of oil used in caramelising should not be too much.
- Shuǐ-Yóu ba (Water and Oil Mixed caramelising): A small portion of oil is added first to grease the wok and then appropriate amount of water is poured in. Bring it to a boil, add sugar in, simmer it in medium to low heat. Caramel made with this method is light golden in colour with long and crunchy sugar threads.

The general process of Basi (Caramelising) technique shown as below:

Selecting ingredient → preparing and/or cutting → wet or dry corn starch coating (or not) → oil blanching → making caramel → coat ingredient with the liquid sugar and pull the sugar threads → place in a pre-greased plate → serving (with a bowl of water)

Key skills

- Basi dishes should always use fresh fruit, which should be peeled and stoned to avoid colour change. The ingredients are usually cut into strips, chunks, or balls.
- Most ingredients should be wet coated and then oil blanched. Occasionally, oil blanching is done without coating. A few coating options are available including egg white, eggs, crispy batter, day cornflour. Some ingredients will need to be steamed until soft and tender, and then mashed and shaped into strips or balls prior to coated oil blanching.
- Chefs should be able to create a range of textures such as crunchy from the outside and tender from the inside, crunchy from the outside and crispy from the inside, crumble from the outside and crispy from the inside by using the appropriate coating and ingredients.
- Use a clean wok, add sugar, water or/and an appropriate amount of oil depending on which caramelising technique is chosen. Bring the liquid to a boil, simmer until the syrup turns light brown. Plunge ingredients in, stir well for an even coating on all sides. Remove from the wok, place on a pre-greased plate, serve immediately with a bowl of cool water.
- Make sure the sugar, water and/or oil are well proportioned according to the chosen caramelising technique. Skilful control of the heat during caramelising process is also required to avoid recrystallisation and burning the sugar. Recrystallisation happens when ingredients are mixed with syrup that is not completely liquidised, normally caused by using a lower level of heat.
- Frying the main ingredient and making syrup should be done instantaneously. If the main ingredient is cooked first, it will be cold when the syrup is ready. If mixed with cold ingredients, the syrup gets hardened quickly and reduces the chance to create sugar threads.
- After frying, the main ingredient should be drained well to allow for even coating when mixed with syrup.
- Basi dishes should be served on a pre-greased plate to avoid sticking.
- Basi dishes should be served immediately to create the best sugar threads effect. If the room temperature is low, warmer should be used to avoid syrup becoming solidified quickly.

Features of Basi dishes

Basi dishes have a translucent crystal crust, crispy from the outside and tender from the inside, sweet and aromatic. The long and thin sugar threads are a unique feature.

Dishes cooked with this method

    Basi Banana

    Basi Grapes

    Basi Apple

    Basi Sweet Potatoes

    Basi Honeydew

    Basi Ice Cream

1.18 Ju (Baking)

Ju (Baking) is a traditional Chinese cooking method. It requires to place food in a sealed container on heat. This process will vaporise water within the food to create hot air to cook the food.

There are Stove/oven Baking and Salt baking.

- Stove/oven baking: ingredient is par cooked using another method such as stir-frying or braising, then to be placed in a sealed pot and cooked rapidly over high heat on stove top or in an oven. It requires a high cooking temperature and fast heat transfer to enable the food to release its aroma.
- Yan Ju (Salt baking): ingredient is marinated, wrapped, and then buried in hot salt until cooked.

The process of Yan Ju (Salt Baking) includes the following key steps:

Selecting ingredients → initial processing → marinating and wrapping → baking in hot salt in a pot → cutting → plating (sometime topping sauce is required)

Notes

- In selecting ingredients: either small or large sized ingredients can be used. They should be fresh and tender. Ingredients can be cut into chops, chunks or kept in their original forms.
- Marinating: marinating time should be in accordance with the properties of the ingredients.
- When baking in hot salt in a pot:
  - Although rock salt is preferred, table salt can be an alternative.
  - Ingredient(s) must be wrapped well.
  - The extent to which ingredients are cooked should be kept at "just cooked". The roasting time needed to achieve the "just cooked" effect is determined by the properties of the ingredients.
  - Salt should be evenly distributed.

Key skills

- The ingredients selected should be tender, rich in umami substance, and easy for cooking.
- The wrapping material chosen should be heat resilient. Ingredients must be wrapped tightly to avoid salt getting in.
- The wrapped ingredients should be completely submerged in the hot salt to enable an even heating.
- Those ingredients that are not so easily cooked should be cooked in the hot salt in the pot on low heat.
- The stove or oven temperature should be pre-heat between 150°C and 180°C.

Features of Ju dishes

    Soft and fall-apart tender, umami and aromatic.

Dishes cooked with this method

    Baked Fish Sausage

    Baked Squab with Rose

    Ancient Style Baked Perch

Chinese Culinary Techniques

Section 2 Chinese Cold Dish Cooking Techniques

2.1 Ban (Tossing)

Ban, or tossing, invloves cutting cooked food (when it's cooled down) or raw food into shreds, dices, slices, or strips, and mixing them with seasonings.

Many ingredients can be used to create a wide range of tossed cold dishes with a variety of flavours. Tossed cold dishes (similar to salad dishes) are fresh, tender, and crispy with a clean and refreshing taste. They should be served immediately to avoid from drying out.

Depending on the ingredients used, there are three types.

Types of Tossed Dishes	Notes	Sample Dishes
Raw ingredients	<ul><li>Should use fresh, crunchy and tender vegetables or other ingredients with a similar texture that can be eaten directly.</li><li>Ingredients should be washed and cut before seasoning.</li><li>Some vegetables with over powering flavours could be soaked in ice water or marinated with salt to reduce this effect</li></ul>	<ul><li>Bitter Gourd in Plum Sauce</li><li>Curly Endive in Sesame Sauce</li><li>Spicy Coriander Salad</li><li>Marrow Shreds Salad</li><li>Sweet and Sour Radish Shreds</li><li>Assorted Veggies Salad</li></ul>
Cooked ingrdients	<ul><li>Ingredient can be cut into desired shapes either before or after it is cooked, and then mix with seasoning sauce.</li><li>Water-blanching, oil blanching, boiling are normally used to cook the ingredients. Regardless of the method used, food should be cooked to the level of doness called "just cooked".</li><li>When use water-blanching and boilding methods, always place ingredients in boiling water, remove when they are just cooked, mix with seasoning sauce when the food is hot in temperature to allow better flavour infusion.</li><li>"Just cooked" ingredients could also be rinsed in iced water or a blast chiller piror to seasoning to maintain their crunchy texture and vibrant colours</li></ul>	<ul><li>Beef Tripe in Shacha Sauce</li><li>Razor Clams</li><li>Duck Paw in horseredish oil</li><li>Potato sheds salad</li><li>Eight-Treasure Spinach</li></ul>

171

Part Three

Types of Tossed Dishes	Notes	Sample Dishes
Cooked and raw ingredients	• Make sure the cooked ingredients are cooled down completely before mixing with the raw ingredients.	• Chicken Shreds and Sweet Potato Noodles • Three Shreds • Clams and Cucumber

2.2 Qiang

Qiang is a widely applied cold dish cooking method. Ingredients are often shredded, sliced, or diced, then blanched in boiling water or fried briefly in hot oil, and finally mixed with spicy and aromatic seasoning sauce containing Sichuan peppercorn oil.

Qiang dishes require careful cutting skills. These savoury and aromatic dishes tend to have bright colours, a tender and crispy texture.

The main seasonings used for Qiang dishes are Sichuan peppercorn oil, spring onions, ginger, garlic, fragrant vinegar, curry powder, Chinese mustard, chilli pepper, and white or black pepper.

Qiang dishes can be done in three different ways: Qiang with Water, Qiang with Oil, and Sheng Qiang (Qiang raw ingredients).

Types of Qiang Dishes	Process	Notes	Sample Dishes
Qiang with Water	Qiang with Water is a process where pre-cut ingredients are briefly cooked in boiling water and then immediately plunged into cold water to cool the ingredients down. The ingredients are then seasoned with herbs and spices	Tender, crunchy and less juicy ingredients are suitable for the Qiang with Water method. The blanching time should be brief to ensure the ingredients are "just cooked". The water used for blanching must be boiling	• Qiang Organic Cauliflower • Warm Qiang Seafood • Qiang Kidney Slices • Ningbo Style Jellyfish • Kaiyang Qiang Celery
Qiang with Oil	Qiang with Oil involves dipping pre-cut ingredients into a batter to create an even coating, quick-frying in hot oil, rinsing under running warm water, draining the oil and water out, and then finally adding in seasonings	As this process often uses tender meat, the temperature of the oil and time taken to quick-fry are critical to preserve the smooth and tender texture of the ingredients	• Fish Slices with Pickled Cucumber • Qiang Shrimp with Mushroom
Qiang raw ingredients	Sheng Qiang, or Qiang raw ingredients, is a cooking method to cure fresh raw seafood or fish in liquor and vinegar sauce with salt, minced garlic and minced ginger	Sheng Qiang dishes are not cooked by heat but by alcohol, acid and garlic. Hence, Sheng Qiang dishes must be prepared and consumed with care to minimize the risk of food poisoning	• Qiang Shrimp with Fermented Bean Curd • Qiang Fresh Tuna with Chinese Mustard Sauce

2.3 Yan (Marinating)

Yan refers to a cooking method in which ingredients are prepared with salt, rice wine sauce or sugar and left to infuse in the marination for a period of time in a cool place to absorb the flavour.

Chinese Culinary Techniques

Yan dishes are tender and aromatic with unique flavours.
There is a wide range of Yan dishes. Based on the seasonings used, there are Yan with salt and seasonings, and Yan with sugar and seasonings.

Types of Yan Dishes	Process	Sample Dishes
Yan with salt and seasonings	Yan with salt and seasonings involves marinating the ingredients with salt to remove excessive moisture and/or *astringent taste, and to add flavour. This method also commonly used to prepare the ingredients prior to other cooking methods	• Slightly Marinated Chinese Green Radish • Cucumber in Soya Sauce
Yan with sugar and seasonings	The process starts with adding a small amount of salt to the ingredient, leaving it to infuse for a while, then gently squeese out extracted liquid, mix the ingredient(s) with white sugar and other seasonings for further marinating to add flavours	• Five-flavour Cucumber Strips • Shanghang Style Chinese White Radish • Asparagus Lettuce (also known as Celtuce or Chinese Lettuce) in Red chilli oil • Sweet and Sour Radish.
Notes		
• The time required to achieve the best Yan (Marinating) results depends on the texture and size of the ingredients, and the amount of salt use. • Suitable ingredients for Yan method are tender and crunchy seasonal fruits and vegetables. • When seasoning, sweet and sour tastes should be in balance		

*"Astringent" is used to describe a unique sharp, slightly acidic, or bitter taste of some vegetables or fruit when they are consumed in raw. In Chinese language, this taste is referred to as 涩 (se).

2.4 Pao (Pickling)

Pickling is a cooking method that preserves ingredients in a liquid based marinade, usually contaning vinegar, alcohol or brine.
Pickled dishes are tender, crunchy, refreshing with an integrated taste of saltiness, sourness, spiciness and sweetness.
Depending on the ingredients used and the flavour of marinade, there are two types of Pao (Pickling): Salty Pao and Sweet Pao.

Types of Pao Dishes	Process	Sample Dishes
Salty Pao	Salted Pao (Pickling) mainly uses salt, liquor, Sichuan peppercorn, dried chilli, pickled chilli, ginger, garlic, sugar. Salted Pao dishes give a combined salty, spicy, and sour taste. The acidic effect is a result of fermentation	• Sichuan Pickled Vegetables • Phoenix Paw in Pickled Chillis
Sweet Pao	The main seasonings of Sweet Pao are sugar and white vinegar. It tastes sweet and sour	• Curry Flavoured Veggi • Pheonix Paw in Thai Style Sauce • Lotus Root in Orange Spuash • White Gourd in Orange Juice

173

Notes

- Vegetables and fruits should be fresh, crunchy, tender, and low in fibre. Ingredients should be carefully cut into desirable shapes.
- To process Salted Pao, a Chinese style pickling jar is required. The jar has two lids. The smaller lid goes into the bigger lid. When the bigger lid is filled with water, it creates an air-tight seal. When pickling, the jar should be stored in a cool place, water in the bigger lid should be changed daily or once every two days. Ingredients and utensils used should be totally grease free to avoid food spoilage.
- Pickling brine should be kept clean. Never use bare hands or greasy utensils to fetch the pickled ingredients out of the jar.
- Pickling time depends on the season, the maturity (newly prepared or aged), tastes (rich or light, salty or sweet) of the brine.
- Brine can be reused for pickling if there is no spoilage. Please remember to remove all the pickled ingredients prior to adding new ones in. More seasonings should be added to keep the desirable flavours.
- Most of the Sweet Pao dishes are required to be stored in the fridge while preparing and it normally takes a day to be ready for serving

2.5 Boiling in plain water

Boiling in plain water is a cooking method wherein ingredients are initially cooked in plain water on high heat, bring water to the boil, reduce the heat to low or switch it off, continue with the cooking in hot water until food attains required doneness, remove from water, once the food cools down, cut it into desired shapes, finally mix with sauce before serving or serve with dipping sauce on side.

Notes

Ingredients should be fully immersed in water when cooking and stir every now and then to ensure it is evenly cooked.

Dishes cooked with this method

- Chopped Cold Chicken with Ginger and Spring onion
- Sliced Cold Pork with Minced Garlic Sauce
- Mala Chicken
- Guaiwei Pig's Tongue
- Sliced Cold Lamb

2.6 Boiling in salted water

Boiling in salted water is a cooking process wherein ingredients are marinated first and then boiled in salted water. Alternatively, ingredients could be boiled in salted water first and then marinated.
Salted water boiled dishes are savoury, umami, light, tender, and refreshing.

Notes

- Ingredients should be fully immersed in water when cooking.
- Large and tough ingredients should be soaked in cold water or blanched prior to Yan (marinating) to remove overpowering flavours. It should be cooked in cold water with seasonings on high heat, when water starts to boil, reduce the heat to low, cook until food is done. Avoid boiling on high heat for too long. It will affect the texture of the ingredients.
- Sugar and spices that will add colour to the ingredients are avoided whilst cooking. If the dish is featured with tender texture and umami taste, it is not advised to add salt early in the cooking process as the electrolytes in salt can cause coagulation of proteins ingredients. This means food will require

a longer time to be done and the texture of it tends to be tougher.

Dishes cooked with this method
- Salted Nuts
- Salted Edamame (green soy beans in pods)
- Salted King Prawns
- Salted Gizzards
- Salted Beef Shin
- Nanjin Salted Duck
- Salted Pork Shank

2.7 Lu

Lu is a slow cooking method. It requires to blanch ingredient in water or oil briefly first, and then slow cook the food in flavourful sauce over medium to low heat for an extended period of time.

Meat and poultry ingredients are the best choice for Lu dishes. Soy food can be a plant-based alternative. Most of Lu food is preferred to be kept in its broth before serving to achieve an enhanced aromatic and umami taste. Some Lu dishes require the ingredients to be removed from the broth once it is cooked. A thin layer of oil will be brushed on to the food to keep it from drying or discolouring.

Lu dishes often have a glossy red brown colour, an umami, pure and rich, and aromatic flavour with a moist texture. Depending on the seasoning sauce used, there are two different types of Lu dishes.

Types of Lu Dishes	Main Seasonings	Sample Dishes
Red Lu, or red broth	Light soy sauce, dark soy sauce, red yeast rice, sugar, cooking wine, vegetables with aroma, and spices	• Lu Baby Pigeon • Lu Poussin Chicken • Lu Tofu • Lu Goose • Lu Duck Gizzards • Lu Beef Tripes • Lu Beijing Duck
White Lu, orwhite broth	Seasonings with colours and sugar are excluded	• Lu Duck Tongue • Lu Duck Palm • Lu Spring Bamboo Shoot

Notes
- Herbs and spices used in the Lu method should be tied in cheesecloth or muslin food bags.
- If preserved well, the broth can be used repeatedly. Small amounts of stock, herbs, spices and other seasonings can be added to the broth each time after use to maintain the flavour. Long term preserved broth, known as the "aged broth", provides dishes with an intense taste and aroma.
- The best way to preserve "aged broth" is to bring it to a boil after each use, remove any white foam and oil floating on the surface, filter out leftover food from previous cooking and add a desirable amount of stock and seasonings in to keep the intensity of the flavour

Good practise note: blast chill the broth after use.
- Herbs and spices high in oxalic acid such as star anise and cinnamon should be avoided in Clear Lu Cao-guo Cardamom (sometimes spelt Tsao-ko Cardamom), Bai Zhi (White Angelica Root) and Sichuan peppercorns are good substitutes. The broth should be preserved in stainless steel container, not a cast iron one.
- Heat control needs to be carefully handled in the Lu process. Typically, it starts with bringing the broth to boil on a high heat, and finishe with simmering the ingredients on low heat. When cooking a variety of ingredients together, the cooking time for each ingredient will be different based on its texture.
- Always use clean utensil to fetch cooked ingredients to prevent the broth from being spoiled

2.8 Jiang

Jiang is a slow cooking method in which ingredient is marinated, water blanched or oil blanched, and then cooked in savoury liquid seasoned mainly with fermented flour paste or soy sauce. Typically, Jiang requires using high heat at the beginning and reducing to medium or low heat to cook the food until it achieves the desired texture and colour.

Jiang dishes are moist, savoury, tender, and aromatic with a glossy dark red-brown colour.

Cooking process of Jiang and that of Lu are very similar so much so chefs tend to use a combined term "Jiang Lu" to refer to them. However, there are three steps that set Jiang and Lu apart as shown in the table below:

Steps	Jiang (Stewing)	Lu (Braising)
Marinating	✓	✗
Use of the "aged broth"	✗	✓
Thickening the broth at the end of the cooking	✓	✗

In Jiang process, ingredient is marinated first and then cooked in savoury broth while Lu process doesn't involve marination.

In Jiang process, the broth is freshly made and consumed with the food. In Lu process, broth is kept carefully for future cooking. To use "aged broth" is a means of enhancing the flavour and aroma of food.

Jiang process is finished with thickening the broth on medium heat. Lu method doesn't require this step. As a result, the colour of Jiang dishes is darker than that of Lu dishes.

Notes

- Commonly used ingredients in Jiang method include beef, pork and poultry.
- To ensure the ingredients is evenly cooked and coloured, the following steps need to be followed: Bring the flavoured liquid to a boil, plunge in marinated ingredients, reduce to low heat for simmering, and stir twice during the process.
- The cooking time is determined by the texture and size of the ingredients. When 30% of the cooking time is left, increase to medium heat to thicken the broth to enhance the flavour and colour of the food.

Dishes cooked with this method

- Jiang Beef Shank
- Five-spice Jiang Chicken
- Jiang Squab
- Jiang Pork Ribs
- Jiang Pork Trotters

2.9 Dong (Aspic)

Dong method involves boiling, steaming, and then filtering gelatine-rich ingredient such as agar or pork skin to make a thick and sticky soup that can be jellified. Cook the main ingredient separately, and then mix it with the soup. Leave in the fridge to gelatinise.

Aspic dishes are translucent, tender, smooth, and refreshing. They are often presented in an artistic pattern. The translucent jellified soup is called crystal. The name of a aspic dish is normally beginning with "crystal" and followed by the main ingredient.

The choice of ingredients is seasonal. Less fatty animal ingredients such as chicken and prawns are preferred in summer while pig trotter and mutton are more popular in wintery months.

Types of Dong Dishes	Main ingredients and flavour of the aspic	Sample Dishes
Salted Dong	Animal ingredients, pork skin, salted taste	• Crystal Chicken • Crystal Shrimp • Crystal Three Treasures • Crystal Pork Shank • Crystal Mutton • Crystal Duck Tongue • Crystal Duck Strip
Sweet Dong	Fruits, vegetarian options of gelatine rich ingredient such as agar agar powder, sweet taste	• Watermelon Aspic • Assorted Fruits Aspic
Notes • Gelatine-rich ingredients include pork rind, tendons, and fish skin. Alternatives include agar agar powder and edible gelatine. • To make the gelatine, the amount of water used should be in proportion so that the broth would not be too thick or too thin. • Tender animal proteins are the best choice for making Dong dishes. • Make sure the broth is cooled down completely before mixing it with the main ingredient. Hot broth will affect the colour of the main ingredient. • The main ingredients should be totally immersed in the broth. • Ingredients for garnishing should be bright in colour		

2.10 Frying – Lu

Frying-Lu is a combined cooking method. It starts with deep-frying the ingredients. Once this is done, the food will be further processed with one of the below methods:
- Simmered over medium or low heat in sauce
- Topped with freshly made sauce
- Marinated in sauce

Therefore, Frying-Lu method can be further divided into frying and simmering, frying and topping, and frying and steeping

Fry-Lu dishes are light brown in colour, crispy and dry with an aromatic and rich flavour.

Types of Frying-Lu Dishes	Cooking Method	Sample Dishes
Frying and simmering	Ingredient is fried first, and then simmered in liquid on low heat and finally use medium heat to slightly thicken the liquid	• Five-fragrant White Carp • Tianjin Style Crispy Hairtail
Frying and topping	Ingredient is fried first and set aside, cook the topping sauce, pour the sauce on to the surface of the food when the sauce is still hot	• Sweet and Sour Crispy Fish • Wuxi Crispy Eel
Frying and steeping	Ingredient is fried first, cook the steeping sauce separately, immerse the food in the sauce to absorb the flavour	• Aromatic Chub Mala Crayfish

Notes
- Generally speaking, batter is not required, ingredients are fried in oil at 160 degrees. Large cuts of meat may be fried twice to be crispy on the outside and tender and moist on the inside.
- If marination is required before frying, remember not to add too much salt in the marinade. Marinated ingredient(s) should be drained well before frying to avoid splashing.
- The amount of steeping sauce should be in proportion with the food. The required time for steeping depends on the texture of food.
- In terms of heat control, simmering in broth over low heat is used to allow the food to absorb more flavour, boiling on medium heat for a short time is used to reduce the broth

2.11 Hot Qiang

Hot Qiang is a cold dish cooking process wherein ingredient is scalded in boiling water, removed after a brief and timed blanching, and seasoned immediately.

There is a variety of flavour combinations in Hot Qiang, which can be roughly divided into original flavour and spicy flavour. Essential seasonings used to create the original flavour include soy sauce, salt, sugar, vinegar. Spicy flavour is created by Sichuan pepper, chilli pepper, spring onions, ginger, garlic and cooking wine.

When seasoning, cooked ingredients should be mixed with sauce immediately. In some cases, hot oil topping is required.

Notes
- Hot Qiang requires very brief blanching in boiling water hence the selection of ingredients should be those of extremely tender and crunchy texture and very fresh.
- Ingredient should be finely sliced or shredded for easy cooking and better taste.
- The cooking time is crucial to ensure a tender texture. Ingredient should be "just cooked".

Differences between Hot Qiang method and Qiang method

Qiang is the collective name of a few methods in which water blanching or oil blanching is the main cooking method. Comparing to other variations of Qiang, Hot Qiang is unique in the following three ways:

1. Selection of ingredients: only use very tender, crunchy and fresh ingredients.
2. Process of blanching: no need to shocking or refreshing in cold water after scalded in boiling water.
3. The time required for blanching is the shortest, between 10 and 20 seconds, due to the texture of ingredients.

Dishes cooking with this method
- Hot Qiang Baby Cuttlefish
- Hot Qiang Asparagus
- Hot Qiang Beef Tripe
- Hot Qiang Sea Snail Slices

2.12 Liu Li (Caramelising)

Liu li method is used to cook dessert. It involves making caramel and dipping ingredients in it to form an even sugar coating. When the sugar coating hardens, it becomes a transparent light brown crust which looks like amber or a traditional Chinese glass art form - Liu Li.

Liu li dishes are sweet with crusty caramel coating. People often call this type of dishes the Liu Li desserts.

Notes
- Liu Li caramel is ready when you can easily pull it into sugar threads.

- When the ingredients are coated, put them onto an oiled plate immediately and separate them to avoid sticking together. Cool them in a well-ventilated place and serve when the caramel crust hardens.

Dishes cooked with this method
- Liu Li Nuts
- Liu Li Lotus Seeds

2.13 Zao

With Zao method, ingredient would be boiled first and then bathed in a mixture of salt and zao marinade. Zao marinade is a traditional Chinese seasoning sauce made of Jiu Zao (Chinese rice wine lees) and salt (100∶4).

Jiu Zao (酒糟) is the lees left over from Chinese rice wine production, contains ≥3% of alcohol and high nutritious values, provides health benefits. It is often used as a marinade base in Chinese cuisine to remove fishy or gamey odour, enhance the umami flavour, and increase appetite.

Recipe for making Zao marinade differs from region to region. A typical method is to add salt, spring onions and ginger to broth and bring it to a boil. When it cools down, add red lees or fragrant lees cube in it. Filter the broth, and then add a small amount of alcohol. Ready-to- use Zao marinade is commercially available.

Zao dishes have a unique umami flavour and a tender texture.

Notes
- Fresh white meat is a typical ingredient of Zao dish. There are vegetarian choices as well. To highlight the aroma of Zao marinade, ingredients with strong flavour should be avoided.
- Zao dish doesn't require a prolonged cooking time. Ideally, Ingredient should be just cooked.
- Zao dishes are best to be served chilled (under 10°C).

Section 3 Chinese Pastry Making Techniques

The Chinese term Mian Dian (面点), commonly translated as Chinese pastry, refers to all types of flour-based food that is made of a dough consisting of flour (wheat, rice, other grains, beans), water, or raising agent (physical, biological, or chemical), or shortening agent (fat). The flour dough may be plain, savoury or sweetened, with or without a savoury or sweet filling (made by using a wide range of ingredients including vegetable, fruit, river fish/shrimp, seafood, meat, poultry, and eggs).

Chinese Mian Dian can be cooked with a variety of Chinese cooking methods such as deep frying, frying, steaming, baking, poaching, boiling etc.

Part Three

Although it might be known as Chinese pastry in the English-speaking countries, Mian Dian is different from the Western concept of pastry as it includes buns (Chinese bread), noodles, dumplings, and flatbreads.
Chinese Pastry module includes the learning of two key types of techniques, namely dough making techniques and filling making techniques.

3.1 Flavours and Styles of Chinese Pastry

Chinese pastries vary in flavour and each province has its regional style. Nevertheless, it is widely accepted by gourmets and industry experts in China that there are mainly two flavours: the "southern flavour" and the "northern flavour" and among all of the reginal styles, the top three are "Cantonese style", "Jiangsu style" and "Beijing style".

3.1.1 Beijing Style

Beijing-style or simply Jing-style pastry generally refers to the pastry food that are produced and consumed in northern China (north of the Yellow River) including North China region and the Northeast region. This style is best represented by Beijing area hence the name.

Famous varieties include hand-pulled noodles, yipin shaobing, vegetable oil flatbread, duyichu shaomai (dumpling with an open mouth), goubuli baozi, shaobing with minced meat filling (a royal cuisine pastry dish), thousand-layered cake, aiwowo (steamed rice cake with sweet stuffing), wandouhuang (powdered sweat pea cake), etc.

Beijing-style pastry originated in countryside in North China, Shandong, the Northeast where there were diversified population of Han, Manchu, Mongolia, Hui (Chinese Muslim) and other ethnic groups. During its formation, Jing-style pastry drew on the best practice of the pastries in different regions and of different ethnic groups and was also influenced by pastries in the south and the Royal cuisine style. Jing-style pastry has the following characteristics.

1. A wide variety of ingredients. Jing-style pastry dishes use wheat, rice, beans, foxtail millet, broomcorn millet, eggs, milk, fruits, leafy and root vegetables as the main ingredients accompanied by minor ingredients and seasonings as wheat flour being the most commonly used main ingredient.

2. A wide range of pastry varieties. Each variety would include a large collection of various food. For example, Shanxi noodles include cheng mian (hand-pulled noodles), daoxiao mian (knife-sliced noodles), xiaodao mian (thin egg noodles), boyu mian (little fish-shaped noodles); Beijing street food include cut cakes, soy milk congee, Tangcha, Miancha, hawthorn gelatine, caramelised fried dough, shaobing, aiwowo, wonton soup, sticky rice ball soup, etc.

3. Sophisticated pastry making techniques. Jing-style is particularly good at making wheat flour dough food, soup/gravy, and savoury meat fillings. Seasoned with spring onions, ginger, soybean paste, flavour enhancer, and sesame oil, Jing-style meat fillings are prepared by stirring meat ingredient in one direction while gradually adding in a good quantity of water or broth to make the meat filling light and sticky. When cooked, it would have a tender and soft texture and a savoury and umami taste. There are various unique techniques to make dough food, for example to make the noodle threads vegetable oil flatbread involves pull flour dough to fine noodle threads first, coil them up, flatten the dough with a rolling pin or palm, and then fry; poria mushroom crêpe is as thin as paper, mung bean pancake is as thin as the wing of a cicada.

4. Integrated flavours of different ethic groups of China include Han, Mongolia, Hui (Chinese Muslim), and Manchu with the legacy of the Royal cuisine.

3.1.2 Jiangsu Style

Originated in Suzhou city and Yangzhou city of Jiangsu Province, Jiangsu style or simply Su style is popular in the lower reaches of the Yangtze River covering Jiangsu Province and Zhejiang Province. The famous pastry food includes Baozi with three-dice filling, Jade shaomai, Sticky rice cakes, Boat dim sum, Tangbao (soup buns), Zongzi. Sustyle pastry making techniques emphasise equally on the colour, aroma, taste, and shape of the food. Su style pastry food tends to have vibrant colours, savoury in taste with a hint of sweetness. The most well-known feature is the use of dong (aspic) in Baozi filling preparation to make the filling soupy once cooked.

1. Sophisticated styles with great varieties. Sustyle pastry can be further divided into Suxi style, Huaiyang style, Ninghu styles, Zhengfen style, each has a great variety in terms of pastry dishes.
2. Meticulous shaping techniques to create the "small but exquisite" appearance. Sustyle pastry dishes are shaped with meticulous care. For example, the boat dim sum is regarded as an artwork for its diversified lifelike shapes of birds, animals, shrimps, fishes, insects, fruits, and flowers in vivid colours.
3. Strict on the selection of ingredients. there is a strict specification for each ingredient used to make the Sustyle pastry dishes covering the origin of produce, variety, season etc.
4. Skillful in utilising the natural colour and aroma of ingredients. For example, using roses, sweet osmanthus, jojoba, red beans, artemisia (wormwood) as natural colouring to add colour and aroma to pastry food.

3.1.3 Guangdong (Cantonese) Style

Guangdong is on the southeast coast of China. With a mild climate, abundant rainfall, it is an ideal land for rice farming. Without surprise, most of the local dim sum food is made of rice, such as Lunjiao fermented rice cake, radish cake, glutinous rice cake, fried rice cake etc.

Cantonesestyle pastry is very unique comparing to Jingstyle and Sustyle in terms of its doughs and fillings. In the past 100 years, it absorbed some western pastry making techniques to have diversified and enriched its pasty varieties. With its international influence, dim sum, a general name of Cantonese flour-based dishes, is arguably the most well-known Chinese pastry food in the world. Cantonesestyle pastry has the following four characteristics:

1. A wide variety of doughs that are made of traditional and common ingredients (rice, rice flour, or wheat flour etc.) as well as extraordinary and local ingredients (pop rice, dried fruit flour etc.), water, oil, sugar, eggs.
2. A wide range of fillings and flavours that are created by using meat, seafood, grains, vegetables, seasonal fruits, dried fruits and nuts. For example, char siu bao.
3. Traditional Chinese pastry making techniques mixed with Western pastry making techniques in specially in dough making.
4. Dim sum dishes are offered according to seasons. For the spring, there are steamed dumplings, thin pancakes with bean sprout, and rose-flavoured loquat; for the summer, there are water chestnut cake, dumpling with duck and dried organ peel filling, chilled rice cake with watermelon juice; for the autumn, there are crab roe soup dumpling, autumn lychee taro; and for the winter, there are chicken with glutinous rice with sausage, eight-treasure sticky rice etc.

3.2 Dough Making Techniques

In this session, techniques used to make various wheat flour doughs and glutinous rice flour doughs are introduced.

3.2.1 Cold Water Dough

Cold water dough is made of wheat flour and cold water. It is commonly used to create noodles, wontons, and water boiled dumplings etc. By kneading, the proteins in the flour, gliadin and glutenin, combine to produce gluten. Hence, cold water dough is strong and hard with good extensibility. It is particularly good for making dumpling wrappers and noodles.

To make cold water dough, place flour on a flat work surface, make a well in the middle of the pile, pour water into the well, bring the flour from the sides of the well into the water, keep mixing until wheat ear shaped lumps start to show. Averagely, the flour and water ratio is 2:1 which may need adjustment as flour's water absorption ability may vary.

3.2.2 Warm Water & Hot Water Dough

Dumplings and flatbread can be made with heated water. This has a major effect on the starch in the flour. As we have seen before, flour can have as much as 70% starch. When you make a dough, it is usually made with cold water. Warm liquid activates the gluten to give it a network to hold in the carbon dioxide and makes the dough rise.

Scalding your flour with boiling water begins another chemical reaction which is starch gelatinisation. This is where the starch in the flour absorbs the water instead of letting gluten be formed. This network of gel polysaccharides (heated starches) that are formed and leaves the gliadin and glutenin lacking the water they need to produce strong gluten strands.

By taking water away from the gluten and starting starch gelatinization, instead you produce a dough that is softer in texture.

3.2.3 Leavened Dough

Raising agents, or leaveners, are used in making buns, cakes, and flatbreads. They improve texture and visual appearance by creating air pockets within a dough or batter to give the final product a light and fluffy texture. There are three types of raising agents, namely physical, biological, or chemical.

Carbon dioxide gas is most often responsible for the leavening action in steamed, baked, or fried food and can be produced by biological agents like yeast, or chemical agents such as baking soda and baking powder.

Physical Raising Agent

In Chinese pastry making, the physical raising agent commonly used is air. Air is often incorporated into batters when fat and sugar are creamed together. Briskly whisking fat with sugar traps small pockets of air within the fat. Air can also be used as a leavener when whipping egg whites. In this instance, the air becomes trapped within a protein matrix in the egg whites, causing expansion.

Biological Raising Agent

Yeast is a biological leavener. It is a living organism that ferments sugars for energy and carbon dioxide gas is a by-product of this process of fermentation. To begin the fermentation process, yeast requires carbohydrates and moisture. Warmth speeds up this reaction, although it is still relatively slow. Because yeast produces carbon dioxide at a slow rate, it is often used in buns and flatbreads that have a strong gluten matrix that can hold the gas in for long periods of time. Liquid batters, like those used for pancakes, are too weak to keep gas trapped for that length of time and they need a faster-acting leavener such as baking soda.

Chemical Raising Agent

Two chemical leaveners are baking soda and baking powder. Baking soda is a natural alkaline powder that produces carbon dioxide gas when combined with an acid. Since the reaction occurs rapidly, baking soda is an ideal leavener for soft or weak batters like pancakes, muffins, biscuits, and other quick breads. Vinegar, yogurt, lemon, lime can be used as the acid in this reaction.

Baking powder is similar to baking soda but it already contains the acid necessary to react. The acid in baking powder is in the form of a salt, which means that it will not react until combined with water. Baking powder is an ideal leavened for recipes that do not contain a lot of other acidic ingredients, such as cookies. Most baking powders sold commercially today are double-acting, which means that it will produce gas twice—once when water is added and again when the mixture is exposed to heat. Double acting baking powder provides a consistent and reliable leavening action.

3.2.4 Laminated Dough

Laminated dough is the process of sandwiching dough and fats together and then folding them together to make layers. The fat used in Chinese pastry is usually vegetable oil which acts as flavouring to enrich the dough. The moisture that emits from water in the pastry and vegetable oil is turned into steam and caught between the layering. This is what causes the pastry to rise.

3.2.5 Shortcake Dough

To make this dough, coat the flour with the fat then bread crumbing it together with your hands to fully incorporate it together.

Coating the flour with the fat makes the pastry tender once cooked, with the addition of water this makes the pastry strong due to the gluten development.

Continue to knead the dough by pushing it on the surface away from your body and bringing it back towards you. This action will shorten the gluten strands by covering them with the fat, so the gluten is developed but it has been shortened.

3.2.6 Glutinous Rice Dough

Glutinous rice dough is the name given to the appearance of the sticky rice after it has been steamed and then manipulated to develop the starch within the rice. Rice does not contain proteins that turn into gluten and is therefore gluten free.

To produce the dough, wash the sticky rice and soak in cold water for 30 minutes. Drain the sticky rice and place it in a steamer, steam for 30 minutes until fully cooked. Tip out of the bowl on to a clean wet cloth or cling film and work the rice until it is malleable and forms a dough.

Section 4 Modern Cooking Methods

Thermo circulator – This controls the temperature of the water within a water bath. Place desired items within a vacuumed sealed bag to whatever temperature you require. The end result is one of precision allowing for complete control and consistency. You can cook anything from meat, fish and poultry to fruit.

Paco Jet – Whilst this machine has been readily available for some time, it has changed the way you can make purees, ice creams and sorbets. This allows you to batch freeze your desired product and blend it when needed to a perfect consistency. This is great for dietary requirements and portioning control. Ice creams are blended to an extremely smooth consistency and made with ease.

Thermomix – This blender will turn whole cloves of nutmeg into a fine powder which shows you the power it has. This item can cook and blend at the same time, perfect for chocolate ganaches and creme patisserie. It can also weigh, chop, saute, knead doughs, whip and cook items by steaming with additional add on equipment.

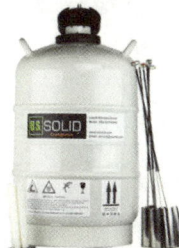
Liquid Nitrogen – This has been used for some time now with fantastic results in the food and drink industries. The N_2 released produces a fantastic smoke that bellows out for a dramatic appearance. It is also used to freeze items such as ice cream. It freezes the water particles so fast that the mouth feels like it has hardly any ice crystals within the ice cream, producing a very smooth product.

Part Three

Dehydrators – This has been a popular kitchen aid for a long time. Producing items that can be blended due to the water being removed from the desired product. This leaves an intense concentrated taste which aids towards your dishes complexity.

Vegetable Sheeter – This has become a popular item over the years and can cut sheets of vegetables which can be used to inspire your creations.

Cream Whipper – Widley is used for multiple ingredients. This gas charged whipper can be used to create hot and cold foams to create another texture and flavour for your dishes with a great appearance.

Spherification – Use Sodium Alginate and calcium to produce modernist caviar and ravioli to add another dimension to your dishes. Also, try reverse Spherification for items containing a higher percentage of calcium to create the desired size sphere that when opened reveals a smooth flow of liquid.

Smoking gun – Over the years, there has been a trend to heighten senses whilst eating. The aid of a smoke gun brings scented smoky flavours to dishes and the item being burnt can even be infused with flavours such as citric fruits and alcohol.

Section 5 Chinese Culinary Vocabulary

We've already spent a good amount of time in the Chinese kitchen, learning the words for kitchen appliances as well as dishes, utensils, and more. Now you've got everything you need, it's time to learn some useful Chinese vocabulary for cooking:

Preparation
- peel (剥 – bō)
- cut (切 – qiē)
- slice (切片 – qiē piàn)
- chop (斩 – zhǎn)
- mince (剁碎 – duò suì)
- cutting board (菜板 – cài bǎn)

Chinese Culinary Techniques

- cleaver (切肉刀 – qiē ròu dāo)
- rolling pin (擀面杖 – gǎn miàn zhàng)
- peeler (削皮器 – xiāo pí qì)
- masher (捣泥器 – dǎo ní qì)
- grater (擦菜板 – cā cài bǎn)
- mixing bowl (搅拌碗 – jiǎo bàn wǎn)
- whisk (打蛋器 – dǎ dàn qì)

Cooking Method Without Heat
- tossing (拌 – bàn)
- marinating (腌 – yān)
- pickling (泡 – pào)

Cooking Method With Heat
- boil (煮 – zhǔ)
- steam (蒸 – zhēng)
- pan-fry (煎 – jiān)
- stir-fry (炒 – chǎo)
- deep fry (炸 – zhá)
- roast (烤 – kǎo)
- bake (烘 – hōng)
- salt baked (盐焗 – yán jū)
- braise (烧 – shāo)
- stew (炖 – dùn)
- smoke (熏 – xūn)
- scalding (烫 – tàng)
- jellifying (冻 – dòng)

Uniquely Chinese Methods
- Red-cooked (红烧 – hóng shāo) a variation of braising
- Lu (卤 – lǔ), a variation of stewing
- Bao (爆 – bào), rapid stir-frying with extremely hot oil over high heat
- Mèn (焖 – mèn), a variation of braising
- Hui (烩 – huì), a variation of braising

The Room
- kitchen (厨房 – chú fáng)
- sink (洗碗池 – xǐ wǎn chí)
- shelf (搁板 – gé bǎn)
- cabinet (厨柜 – chú guì)
- counter (柜台 – guì tái)
- drawer (抽屉 – chōu tì)

Appliances
- stove (火炉 – huǒ lú)
- oven (烤箱 – kǎo xiāng)
- refrigerator (冰箱 – bīng xiāng)
- freezer (冰柜 – bīng guì)

Part Three

- microwave (微波炉 – wēi bō lú)
- dishwasher (洗碗机 – xǐ wǎn jī)
- toaster (烤面包机 – kǎo miàn bāo jī)
- blender (电搅拌器 – diàn jiǎo bàn qì)
- food processor (食物处理机 – shí wù chú lǐ jī)
- coffee machine (咖啡机 – kā fēi jī)
- slow cooker (慢炖锅 – màn dùn guō)
- tea kettle (茶壶 – chá hú)
- rice cooker (电饭锅 – diàn fàn guō)
- can opener (开罐器 – kāi guàn qì)

For Cooking

- pot (锅 – guō)
- frying pan (煎锅 – jiān guō)
- saucepan (平底锅 – píng dǐ guō)
- wok (炒锅 – chǎo guō)
- baking tray (烤盘 – kǎo pán)
- steamer (蒸笼 – zhēng lóng)
- Hand shovel (炒菜铲 – chǎo cài chǎn)
- wooden spoon (木勺 – mù sháo)
- ladle (手勺 – shǒu sháo)

Tableware

- bowl (碗 – wǎn)
- plate (盘子 – pán zi)
- tableware (餐具 – cān jù)
- silverware (银器 – yín qì)
- spoon (勺子 – sháo zi)
- fork (叉子 – chā zi)
- knife (刀子 – dāo zi)
- chopsticks (筷子 – kuài zi)
- cup/glass (杯子 – bēi zi)
- wine glass (酒杯 – jiǔ bēi)

The Great 8 Chinese Cuisine Styles

- Hui (徽菜 – huī cài) – Anhui
- Yue (粤菜 – yuè cài) – Cantonese/Guangdong
- Min (闽菜 – mǐn cài) – Fujian
- Xiang (湘菜 – xiāng cài) – Hunan
- Su (苏菜 – sū cài) – Jiangsu
- Lu (鲁菜 – lǔ cài) – Shandong
- Chuan (川菜 – chuān cài) – Sichuan
- Zhe (浙菜 – zhè cài) – Zhejiang

Popular Chinese Pastries

- shrimp dumplings (虾饺 – xiā jiǎo)
- soup buns (小笼包 – xiǎo lóng bāo)
- pot stickers (锅贴 – guō tiē)

- siu mai (烧卖 – shāo mài)
- spring rolls (春卷 – chūn juǎn)
- rice noodle roll (肠粉 – cháng fěn)
- meatballs (肉丸 – ròu wán)
- steamed pork ribs (蒸排骨 – zhēng pái gǔ)
- BBQ pork/char siu bun (叉烧包 – chā shāo bāo)
- pineapple bun (菠萝包 – bō luó bāo)
- taro cake (芋头糕 – yù tou gāo)
- congee (粥 – zhōu)
- sticky rice (糯米饭 – nuò mǐ fàn)
- egg tart (蛋挞 – dàn tǎ)

Section 6 Allergen

14 most common allergens in Europe

You need to tell your customers if any food products you sell or provide contain any of the main 14 allergens as ingredients.

- celery
- cereals containing gluten – including wheat (such as Spelt and Khorasan), rye, barley, and oats
- crustaceans – such as prawns, crabs and lobsters
- eggs
- fish
- lupin
- milk
- molluscs – such as mussels and oysters
- mustard
- tree nuts – including almonds, hazelnuts, walnuts, brazil nuts, cashews, pecans, pistachios and macadamia nuts
- peanuts
- sesame seeds
- soybeans
- sulphur dioxide and sulphites (if they are at a concentration of more than ten parts per million)

This applies also to the additives, processing aids and any other substances which are present in the final product. For example, sulphites, which are often used to preserve dried fruit, might still be present after the fruit is used to make chutney. If this is the case, you need to declare them.

For further information visit:

https://www.food.gov.uk › files › media › document › top-allergy-types

Part Three

Chapter 10 Recipes

Section 1 Fish & Shellfish Recipes

WARM QIANG SEAFOOD

Main Ingredients
Shrimp meat 150g
Scallop 150g
Squid 200g
Pak Choi 200g
Red sweet pepper 50g

Seasonings
Soy sauce for fish 30g

Light soy sauce 8g
Meijixian soy sauce 8g
Sugar 3g
White vinegar 15g
Spring onions 10g
Ginger 10g
Sichuan peppercorn 4g
Vegetable oil 20g
Sesame oil 2g

Method
1. Devein the shrimp, remove the dark innards from the scallops and then rinse both.
2. Hold the knife in parallel to the chopping board, and flat slice the squid into two pieces from head to tail. Place a piece of squid face down on the board and cut lines from head to tail, 2mm apart, the depth of which is up to two-thirds of the squid. Turn the squid 90 degrees to either side and cut it at a 45 degree angle, the depth of the first cut is two-thirds of the squid, and the second cut is to separate the piece from the squid.
3. Rinse and cut the ginger, spring onions and red sweet pepper into julienne and soak them in a bowl of cold water.
4. Wash the pak choi and halve them lengthwise. Cover the bottom of the wok with water and bring it to the boil. Drop the pak choi in, wait until water is back to boil, remove the vegetable from the liquid and briefly place it in cold water. Drain and lay pak choi on both sides of the plate with the leaves facing inwards.
5. Bring water in the wok back to boil, cook the scallops, shrimp, and pre-cut squid in boiling water for

at least 20 seconds and take them out. Rinse in warm water and drain well.
6. Mix soy sauce for fish, light soy sauce, Meijixian soy sauce, sugar, and white vinegar in a small bowl.
7. Season the seafood with this sauce before placing it on top of the cooked pak choi. Garnish with spring onions, ginger and julienned red sweet pepper.
8. Pour vegetable oil and sesame oil into the wok and drop in the Sichuan peppercorns. The Sichuan peppercorn oil will be ready when the peppercorns turn black and aromatic.
9. Top the dish with Sichuan pepper oil and serve.

Features of the dish

This is a colourful dish with a refreshing taste and strong Sichuan pepper aroma. The tender seafood and crunchy vegetables go well together.

Notes

In terms of the doneness of the food, all ingredients should be "just cooked" to ensure the tender texture.
To prepare the Sichuan peppercorn oil, pour oil into wok, add Sichuan peppercorns in when the oil is still cold, fry on medium heat until the peppercorns turn black and become aromatic.

Part Three

CRYSTAL SHRIMP

Ingredients
Shrimp 500g
Celery 40g
Pork rind 500g
Water 1200g

Seasonings
Salt 25g
Cooking wine 30g
Chopped spring onions 40g
Ginger slices 20g
Sichuan peppercorn 6g
Dried scallop 30g

Method

1. Devein shrimps.
2. Plunge shrimps in the wok with cold water, add 20g spring onions, 10g ginger, 6g Sichuan peppercorn.
3. Bring it to the boil. Cook on low heat for 10 minutes.
4. Remove shrimps from the pot and peel the shell off.
5. Lay shrimps to form a flower shape at the bottom of bowl.
6. Cut celery into brunoise.
7. Leave pork rind in cold water in a wok.
8. Bring it to the boil. Cook for a minute and remove from the wok.
9. Remove fat and impurities, cut the port rind into thin strips.
10. Blanch the pork rind strips briefly remove from the wok and raiseit well.
11. Leave the pork rind in a stainless steel mixing bowl, add 1,200g of water and all the remaining seasonings.
12. Steam for 2 to 3 hours.
13. Filter all ingredients from the broth and pour the broth into the 2 small sized bowls, make sure shrimp is completely submerged in the broth.
14. Garnish with celery cubes.
15. Leave in the chiller for 30 minutes to jellify.
16. Use a small knife to carefully separate the shrimp jelly from the bowl.
17. Removed from the bowl by turning the bowl upside down on the chopping board.
18. Cut the shrimp jelly into three equal portions diagonally.
19. Plating and serve.

Features of the dish
The dish has bright colours. It tastes clean and refreshing.

Chinese Culinary Techniques

FIVE FRAGRANCE POMPRETS

Ingredients
Pomfrets (small sized are perfect) 1,000g
Aniseed starts 8g
Spring onions 50g
Ginger 20g
Garlic 20g
Vegetable oil 4,000g

Seasonings
Cooking wine 28g
Light soy sauce 30g
Salt 10g
Sugar 45g
Five spices powder 15g
Water 250g

Method
1. Remove the fins from a pomfret to make them neat.
2. Remove the gills and soak in water.
3. Repeat the steps until all pomfrets are prepared.
4. Rinse pomfrets under running water and drain well.
5. Pour vegetable oil into the wok, heat it up to 160°C.
6. Plunge pomfrets in, fry for 1 minute. This step is known as "fry the skin".
7. Reduce to low heat, keep the oil temperature at approximately 100 degrees centigrade, continue with the fry for 3 minutes. This step is intended to "cook the meat".
8. Increase to high heat to bring oil to 160°C again, fry for a further 1 minute. This step is to ensure the pomfreds are well cook and crispy.
9. Remove the pomfreds from the wok, drain the excessive oil.
10. Stir fry aniseed stars, spring onions, ginger and garlic with a small amount of oil until aromatic.
11. Add cooking wine, light soy sauce, salt, sugar, five-spice powder and 250g water into the wok, cook until there is a strong aroma coming out of the sauce.
12. Plunge pomfreds in the sauce, cook on low heat for a while and then flip the pomfreds over.
13. Increase to medium heat to reduce the liquid.
14. Remove the pomfreds, slice them, and place on a plate.
15. Decorate the plate with leaves and flowers before serving.

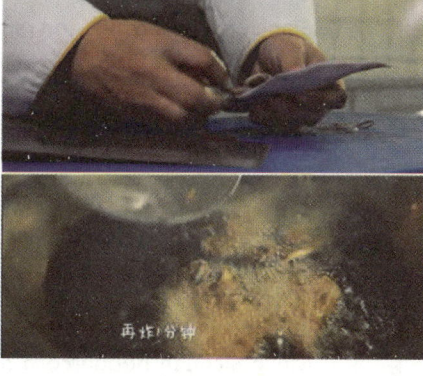

Part Three

HOT QIANG RAZOR CLAMS

Ingredients
Refresh razor clams 1kg

Seasonings
Soy sauce for fish 30g
Light soy sauce 8g
Meijixian soy sauce 8g

Sugar 3g
White vinegar 15g
Shredded spring onions 10g
Shredded ginger 10g
Sichuan peppercorns 4g
Vegetable oil 20g
Sesame oil 2g

Mothod
1. Purge send from razor clams by soaking in cold water for a day.
2. Remove the shells.
3. Bring water to the boil in a wok (cooking wine can be added if wish to), plunge razor clams in.
4. Blanch for 10 seconds and remove.
5. Drain well.
6. Give a firm but gentle press against the colander to squeeze water out.
7. Place razor clams in a mixing bowl, add soy sauce for fish 30g, light soy sauce 8g, Meijixian soy sauce 8g, sugar 3g, and white vinegar 15g, mix them well.
8. Place on a plate, topped with shredded spring onions and ginger.
9. Make Sichuan peppercorn oil by frying 4g in 20g of vegetable oil. Sichuan peppercorn should be added to cold oil.
10. Pour Sichuan peppercorn oil on top of the dish and serve.

Chinese Culinary Techniques

FRIED PRAWN BALLS

Ingredients
Prawns 300g
Fatty 50g
Vegetable oil 1,500g

Seasonings
Egg white 20g

Cooking wine 15g
White pepper powder 1g
Corn flour 25g
Spring onions and ginger infused water 150g
Salt 3g

Method
1. Blend prawns and pork together into paste.
2. Add cooking wine, spring onions and ginger infused water, salt, white pepper powder, egg white, and corn starch.
3. Use your hand to mix the paste well in a mixing bowl.
4. Heat up wok, pour vegetable oil in, heat oil up to 150°C.
5. Use one hand to squeeze out a 2.5cm diameter sized ball and scoop it out with a spoon to plunge into hot oil.
6. Fry on medium heat until prawn balls turn golden and float on the surface.
7. Remove from oil and serve.

Feature of The Dish
This dish is golden in colour, savoury and umami, crispy on the outside and tender on the insider.

Part Three

SOFT FRIED PRAWNS

Ingredients
Prawns 200g
Eggs 2 (egg white only)
Vegetable oil 750g

Seasonings
Cooking wine 2g
Corn starch 25g
Salt 1g
A little bit of flour
Sichuan peppercorn salt 2g (dip)

Method
1. Drain prawns well, marinate with cooking wine, salt and MSG.
2. Make a smooth batter by mixing egg white, corn starch, and flour.
3. Mix prawns well with batter to ensure an even and complete coating.
4. Heat vegetable oil in wok to 150°C, carefully plunging prawns in one by one.
5. Remove when the batter is just cooked.
6. Heat oil up to 180°C, place prawns back to oil, fry until golden.
7. Remove from wok and drain oil.
8. Serve with pepper salt dip.

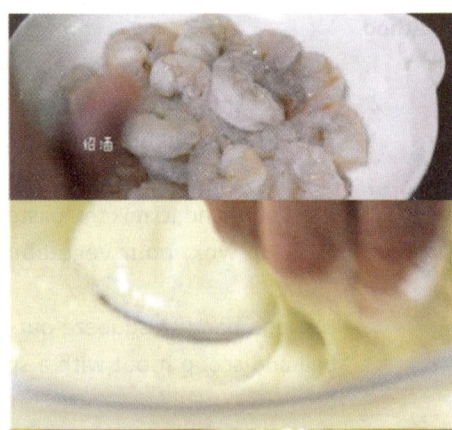

Features of the Dish
Prawns are crispy and soft from outside, fresh, tender, and savoury from inside. This dish is golden in colour.

CRUNCHY BATTER WHITEBAIT

Ingredients
Whitebait 250g
Vegetable oil 1,000g

Seasonings
Salt 2g
Chinese cooking wine 2g
Spring onions 5g
Ginger 5g
Dry flour 150g
Corn starch 50g
Baking powder 3g
White pepper powder 1g
Water 100g
Sichuan pepper salt 1g (dip)

Method
1. Marinate whitebait with cooking wine 2g, salt 2g, and white pepper powder 5g.
2. Mix well dry flour, baking powder, and corn starch, add water in to form batter, add vegetable oil to finish the crispy batter off.
3. Pour oil into wok, heat it up to 150°C, dip a whitebait fish in batter, coated it well before plunge into hot oil. Repeat the process with all the whitebaits.
4. Fry until aromatic and the batter coating raised.
5. Remove to drain oil.
6. Bring oil to 180°C, plunge whitebaits back in.
7. Remove from oil when whitebaits turn golden.
8. Serve with Sichuan Peppercorn salt (dip).

Feature of the dish
Whitebait is tender while the batter coating is crunchy and aromatic.

Part Three

LUCKY FISH FILLET

Ingredients
Fish fillet 500g
Breadcrumb 100g
Vegetable oil 1,000g

Seasonings
Cooking wine 15g
Salt 3g,
Corn starch 45g
Eggs 2
White pepper powder a little bit
Spring onions 5g
Ginger 5g
Catch up sauce (dip)

Method

1. Place a fish fillet on chopping board, poke it with the tip of a cleaver. Repeat this step with the other fish fillet.
2. Seasoning fish fillets in a clean plate with spring onions 5g, ginger slices 5g, and a little bit white pepper powder.
3. Dip fish fillet in corn starch, and then egg, and coated with breadcrumbs. Repeat with the other fish fillet.
4. Bring vegetable oil to 180°C, carefully slide one fillet in, fry until golden.
5. Remove from wok, drain oil well.
6. Cut into strips.
7. Serve with catch up sauce.

Features

Lucky fish fillet is crispy from the outside and tender from the inside with a pleasant golden colour. To impart more flavour marinade overnight.

Chinese Culinary Techniques

FURONG PRAWNS

Ingredients
Prawns 25g
Minced carrot 2g

Seasonings
Egg white 240g

Vegetable oil 1,000g
Salt 5g
Corn starch solution 15g
Dry corn starch 10g
Chicken soup 150g

Method

1. Rinse and drain prawns, add salt 2g, egg white 2g, and dry corn starch in a mixing bowl, mix well.
2. Mix the remaining egg white, salt, chicken broth, corn starch and water mixture in another bowl, ready for use.
3. Add water in wok, bring it to the boil, blanch marinated prawns briefly, remove, and drain.
4. Add prawns in the egg white mixture, mix well.
5. Heat wok up on medium heat, add vegetable oil, pour prawns in, use ladle to stir, keep a sliding movement until egg white gel, remove from wok.
6. Rinse the wok thoroughly.
7. Bring chicken soup 150g to a boil in wok, add in minced carrots 2g, salt 3g, and corn starch water to thicken.
8. Top the dish with chicken soup and serve.

Features of the Dish
The dish is tender, soft, savoury and umami with a light colour.

SOUR FISH STRIPS

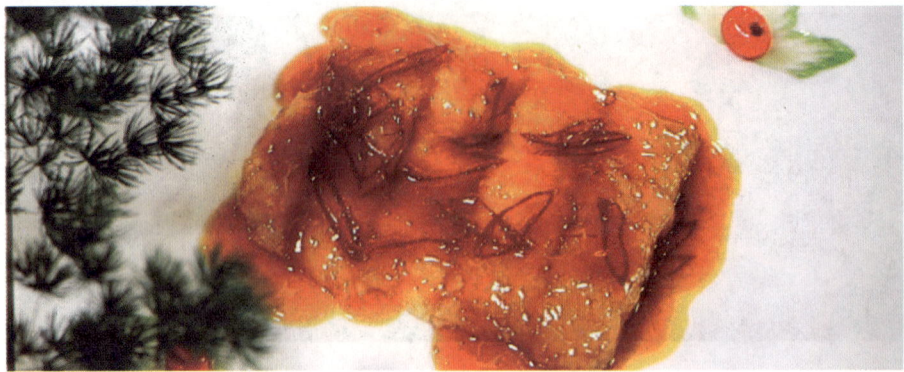

Ingredients
Fish fillet 400g

Seasonings
Cooking wine 15g
Rice vinegar 100g
Salt 3g
Spring onions, ginger, and garlic infused water

Corn flour 50g
Vegetable oil 1,000g
Tomato sauce 10g
Finely sliced dry chili 3g
Sugar 150g
Spring onions and ginger infused water

Method
1. Marinate the fish fillet in cooking wine, spring onions, ginger, garlic infused water, and salt.
2. Roll it in corn starch to form an even coating.
3. Pre-heat the wok, pour vegetable oil in, bring it to 120°C, carefully slide the fish fillet in, fry until crusted and light golden, remove from the hot oil with a colander.
4. Slice the fish fillet obliquely at a 45° angle as shown in the video, place on a serving plate.
5. In the same wok, stir fry tomato sauce, red chili slices, season with cooking wine, vinegar, salt, sugar.
6. Thicken the sauce with Corn starch solution, add a splash of oil to provide a glossy colour to the sauce.
7. Top the fish fillet with the sauce and serve it.

Features of the Dish
The fish is crispy on the outside and tender on the inside topped with a bright glossy red sauce. This dish has a sweet and sour taste with a hint of saltiness and piquancy.

WEST LAKE VINEGAR FISH

Ingredients
A gutted whole grass carp (about 750g-1000g)

Seasonings
White sugar 60g
Cooking wine 25g
Soy sauce 65g
Vinegar 50g
Corn starch 5g
Salt 1g
Chopped spring onions 10g
Sliced ginger 10g
Water 1000g

Garnishing
Minced ginger 5g

Method
1. Butterfly the grass carp, scores on both sides.
2. Add 1000g water together with ginger and spring onions in the wok, bring it to the boil on medium heat, add cooking wine, and slide the fish in.
3. Bring water to boil again, until cooking for about 3 minutes until the fish is just cooked.
4. Remove the fish from the wok, place it on a plate.
5. Leave about 250g of water in the wok, add soy sauce, cooking wine, sugar, salt, corn starch, and vinegar, mix them into a thick sauce. Bring the mixture to the boil.
6. Immediately pour on top of the fish on the plate, garnish with minced ginger, and serve.

Tips: Use a chopstick to poke under the head (closer to the backbone, away from the belly). If the chopstick can go through easily, it means the fish is cooked.

Features of the Dish
This dish has a strong ginger and vinegar aroma and a full body taste starting with sourness, followed by sweetness, and saltiness. The fish is tender and smooth with a flavour akin to crabs.

Part Three

SOFT LIU FISH SLICE

Ingredients
Fish fillet with skin 300g

Seasonings
Salt 3g
Sugar 150g
Cooking wine 25g
Vinegar (Zhejiang style) 100g
White pepper powder 1g
Sichuan pepper oil 3g
Spring onions, ginger, and garlic infused water
Corn starch 50g
Flour 15g
Egg yolk 40g
Vegetable oil 1,000g

Method

1. Slice the fish fillet into a 1cm thick "fish-fan" as shown in the video. Marinate them in cooking wine, salt, white pepper powder, and a little bit of spring onions, ginger, and garlic infused water.
2. In a bowl, mix flour, corn starch, water, egg yolk and a splash of oil to make the batter.
3. Dip the fish-fan one by one in corn starch to have it dry coated.
4. Pour vegetable oil in the pre-heated wok, bring it to 120°C, hold the skin side of the fish-fan, dip it in the batter prior to carefully lower it to the hot oil. Repeat this step with the remaining fish-fans. Fry until they are crushed. Remove from the hot oil.
5. Bring the oil to 210°C, carefully plunge the fish-fans back in, fry until they are crispy on the outside and tender on the inside. Pour the hot oil into the oil drum (oil container) with a colander on, leave fish-fans to drain.
6. Place the wok back on stove, stir in cooking wine, Spring onions, ginger, and garlic infused water, vinegar, sugar, salt, caramel, water. Bring it to the boil, add in corn starch solution and Sichuan pepper oil.
7. Plunge the fish-fans back in. Mix them well with the sauce by rapidly tossing the wok.
8. Dish out and serve.

Features of the Dish
The dish has a mixed colour of yellow and red. Fish-fans are soft, tender, and umami with a sweet and sour taste and a Sichuan peppercorn aroma.

Chinese Culinary Techniques

OIL BAO SQUID ROLLS

Ingredients
Squid 300g
Green and red sweet peppers 50g

Seasonings
Salt 3g
Cooking wine 5g
Corn starch 10g
Clear broth 20g
Vinegar 10g
White pepper powder 3g
Shredded spring onions 5g
Sliced garlic 5g
Vegetable oil 1,000g

Method

1. Trim the squid on both ends, halve it lengthwise, score the surface.
2. Starting from one corner, cut 4/5 into the squid at a 45-degree angle, keep 2mm between each cut.
3. Turn the squid piece around, repeat the pattern from the opposite corner. Finally, cut the squid piece into two halves crosswise.
4. This cutting technique is called wheatear cutting or cross cutting. When cooked the heat causes the squid to curl and gives the appearance of wheat ears on the surface of the squid.
5. As shown in the video, slice the green pepper and red pepper up at a 45-degree angle.

6. Prepare the starchy seasoning sauce by mixing cooking wine, salt, white pepper powder, clear broth, vinegar, Corn starch solution, shredded spring onions, and sliced garlic in a bowl, set aside.
7. Bring water to a boil in the wok (enough to submerge the squid) on medium heat, blanch the squid pieces briefly, remove from heat when they curl into cylinder shape, drain in a colander.
8. Pour oil into the empty wok, bring it to 180C° on medium heat, carefully slide the squid and sweet peppers in, blanch briefly, remove to drain in the colander.
9. Place the wok back on heat, add the oil-blanched ingredients in, mix the starchy seasoning sauce in, rapidly stir the ingredients with the ladle and toss the wok to make sure ingredients are mixed well with the seasoning sauce.
10. Stir in a splash of oil, toss the wok and serve.

Features of the Dish
The squid is tender and firm with a combined flavour of savoury and umami.

Part Three

FRIED COD STEAK

Ingredient
Cod steak on the bone (Darne) 400g

Seasonings
Salt 2g
Cooking wine 15g
Vinegar 10g
White pepper powder 1g

Vegetable Oil 20g
Corn starch 40g
Spring onion and ginger infused water with 1 teaspoon
Spring onions shreds 3g
Ginger shreds 3g
Egg white 2

Method

1. Add cooking wine, spring onions and ginger infused water, salt, and white pepper powder onto the cod steak, gently rub well.
2. Pour vegetable oil into the preheated wok, heat the oil up to 90°C - 120°C on medium to low heat.
3. Thoroughly and evenly cover the cod steak in corn starch.
4. Dip corn starch coated cod steak in egg white.
5. Carefully lower the fish into the wok.
6. Move the wok in circles to spin the fish around inside the wok.
7. Flip side, when the bottom side turns golden.
8. Keep the fish spinning smoothly inside the wok by moving the wok in circles.
9. Add the shredded spring onions and ginger in when both sides of the fish become golden.
10. Add the cooking wine, vinegar, salt, and a small amount of water into the wok, keep moving the wok in circles, then add a splash of oil.
11. Cook until the liquid is nearly evaporated.
12. Dish out, plating, and serve.

Features of the Dish
The dish is golden in colour, crispy from the outside and tender from the inside with an umami and aromatic taste.

Chinese Culinary Techniques

FRIED PRAWN BURGER

Ingredients
Shelled Prawns 300g
Finely chopped water chestnuts (canned) 50g
Lard 50g

Seasonings
Salt 3g

Cooking wine 5g
Vegetable oil 100g
Corn starch 40g
Spring onion and ginger infused water 1 teaspoon
White pepper powder 1g
Egg white 15g

Method
1. Blend the prawns with pork fat into a smooth paste.
2. Add salt, cooking wine, spring onions and ginger infused water, corn starch, white pepper powder, and egg white, mix well, then add the chopped water chestnut to the mixture.
3. Pour vegetable oil into the preheated wok, heat the oil up to 90°C - 120°C on medium to low heat.
4. Make meat balls by squeezing the paste as shown in the video. Carefully plunge the meatball into hot oil with a spoon.
5. Press the meatballs into burgers in the wok with a ladle. Move the wok in circles to keep the mini burgers spinning around inside the wok.
6. Turn over when the bottom side turns golden.
7. Keep mini burgers spinning smoothly inside the wok by move the wok in circles.
8. Remove from wok when the other side turns golden and serve.

Features of the Dish
The prawn burger is golden in colour, crispy from the outside and tender from the inside, light, umami and appetising.

203

PENG PRAWNS

Ingredient
Prawns 350g

Seasonings
Shredded spring onion 5g
Shredded ginger 5g
Sliced garlic 5g

Cooking wine 15g
Soy sauce 20g
Vinegar 10g
Vegetable oil 1,500g
Corn starch 50g
Corn starch solution

Method

1. Remove the antennas and feet of the prawns, devein as shown in the video, halve the prawn crosswise.
2. In a small bowl, make a starchy seasoning sauce by mixing spring onion shreds, ginger shreds, garlic slices, cooking wine, soy sauce, vinegar, and corn starch.
3. Heat the wok up on high heat, pour oil in, bring it to 210°C, roll prawns in corn starch, carefully place them in the hot oil, push them around and keep them separated with a ladle, remove with a skimmer once they are crisp, set side to drain.
4. Increase the oil temperature to 230°C, plunge prawns back in for the second time frying for about 10 second. Pour the hot oil into the oil drum (oil container), leave the prawns to drain in skimmer.
5. Put the empty wok back on a medium heat, place the prawns in, add the starchy seasoning sauce prepared earlier in, toss the wok to mix well.
6. Remove and serve.

Features of the Dish
The dish has a glossy red colour. It is savoury, umami, sour, and aromatic. The prawns are crispy on the outside and tender on the inside.

RED BRAISED FISH

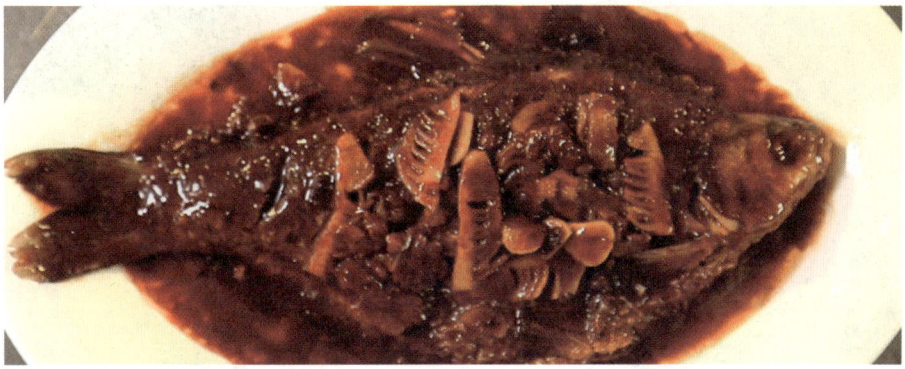

Ingredients
1 whole gutted fish (about 500g-750g, carp or butterfish or alike)
Bamboo shoot slice 50g
Pork slice with fat 50g

Seasonings
Salt 3g
Soy sauce 10g
Cooking wine 5g
Vinegar 15g
Sugar 8g
Corn starch solution 25g
fresh soup 300g
Vegetable oil 1.5kg
Sliced ginger 10g
Chopped spring onions 10g
Sliced garlic 10g
Water 300g

Method
1. Score the fish obliquely on both sides. Leave 4cm between each cut.
2. Pre-heat the wok on medium heat, add oil in, bring it to 210°C, carefully lower the fish in, fry until both sides turn golden. Remove it from the hot oil and set aside.
3. Place the wok back on heat, stir fry the pork slice, bamboo shoot, spring onions, ginger, and garlic with a small amount of oil until aromatic.
4. Lay the fish in, add cooking wine, vinegar, soy sauce, salt, sugar, and water, bring it to the boil on medium heat, cook until the broth is thickened, and the fish is thoroughly cooked. At this point, the flavour of the fish and that of the broth should be combined very well.
5. Stir in corn starch solution to further reduce the broth.
6. Add a splash of oil to the dish to create a glossy appearance.
7. Dish out and serve.

Feature of the Dish
The dish has a glossy persimmon-orange colour, soft and tender.

Part Three

DRY BRAISED CRUCIAN

Ingredients
2 whole gutted crucians
Chopped pork belly 150g
Diced bamboo shoots 150g
Shiitake mushrooms 1cm

Seasonings
Sliced ginger 10g
Spring onion 10g
Garlic slices 10g

Cooking wine 30g
Soy sauce 15g
Vinegar 15g
Pickled red chili pepper chops 20g
Pixian broad bean paste 20g
Salt 1g
Sugar 50g
Water 300g
Sesame oil 50g
Vegetable oil 1,500g

Method

1. Score the fish obliquely on both sides close to the back bone. Leave 1.5cm between each cut.
2. Rub salt on both sides of the two fishes.
3. Pre-heat the wok on medium to high heat, pour the oil in, bring it to 210°C, carefully slide the fishes in, fry until both sides turn golden. Remove from the hot oil and set it aside.
4. Place the wok back on heat, stir fry the pork belly until crisp. Add bamboo shoots, shiitake mushrooms, spring onion, ginger, and garlic with a small amount of oil until aromatic.
5. Add pickled red chili and Pixian broad bean paste, rapidly stir fry.
6. Lay the fish in, add cooking wine, vinegar, soy sauce, salt, sugar, and water, bring it to the boil on medium heat.
7. Reduce to low heat, cook until the broth is thickened and nearly dried out (tip 1), add a splash of oil as the final touch.
8. Dish out and serve.

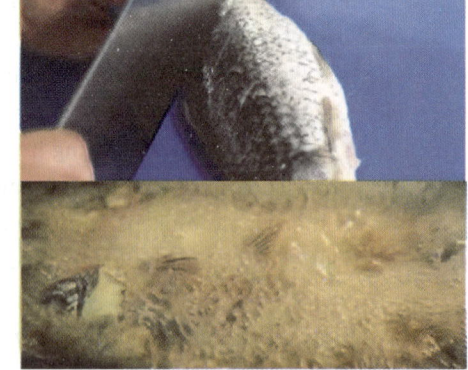

Tips: Use a ladle to scoop some broth onto the upper side of the fish repeatedly to make sure the flavour is evenly infused. Tossing the wok constantly to keep the fish from sticking to the bottom of the wok.

Features of the Dish
Glossy red in colour, the fish is tender and smooth, the pork is crispy and aromatic. There is only a thin layer of oil can be seen on the plate, no liquid.

KAO PRAWNS

![Kao Prawns dish]

Ingredients
Prawns 500g

Seasonings
Vegetable oil 1,500g
Cooking wine 30g
Sugar 75g
Rice vinegar 15g
Finely chopped Spring onion 10g
Finely chopped ginger 10g
Salt 1g
Tomato sauce 30g

Method
1. Remove the antennas and feet of the prawns and devein.
2. Add oil to the wok and bring it to about 170°C on medium heat. Plunge in the prawns and fry until the shell becomes firm, remove them and set aside to drain.
3. Place the empty wok back on medium heat, briefly stir fry the spring onion, ginger and tomato sauce with a small amount of oil.
4. Add prawns in, quickly stir fry them and season with cooking wine, rice vinegar, salt, sugar, and two ladles of water. Bring it to the boil on high heat, remove scum if any, cook until prawns turn slightly red, reduce to low heat.
5. Constantly toss the wok in circles to keep the prawns from getting stuck to the wok.
6. When liquid starts to become thickened, increase the heat level. Keep toss the wok in circles.
7. When liquid is evaporated, drizzle a splash of oil.
8. Dish out and serve.

Features of the Dish
This dish has a rich and pure flavour which can be described as sweet and savoury, umami and aromatic.

Part Three

STEAMED MANDARIN FISH

Ingredients
1 gutted mandarin fish (about 750g)

Seasonings
Sliced spring onion 5g
Sliced ginger 5g
Cooking wine 25g
Salt 2g
pork lard 10g

Soy sauce for fish
White pepper powder 1g
Vegetable oil 20g

For Garnish
Shredded green sweet pepper 10g
Shredded red sweet pepper 10g
Shredded spring onion 8g
Shredded ginger 8g

Method

1. Lay the fish on a chopping board and run a flat blade along the spine, score obliquely on both sides as well, leave a 2.5 cm distance between the cuts.
2. Marinate the fish with salt, white pepper powder, and cook wine for 5 min.
3. Greasy a plate with pork lard, lay a few pieces of sliced spring onions and ginger on it, and place the fish on top.
4. Fill the cavity with a few spring onion and ginger slices, leave a couple on top, and pork lard on.
5. Bring the water in the steamer to the boil on medium heat, place the plate with the fish on to the steamer, steam for 9 minutes with the lid on.
6. Take the plate out from the steamer, remove the spices, drain the liquid.
7. Move to a serving plate, top with soy sauce for fish, garnish with shredded spring onions, ginger, red and green peppers.
8. Finish with a small amount of hot oil and serve.

Features of the Dish
This dish has a delightful assorted-colour, tender and smooth with an umami taste akin to that of crab.

Chinese Culinary Techniques

SALTED PRAWNS

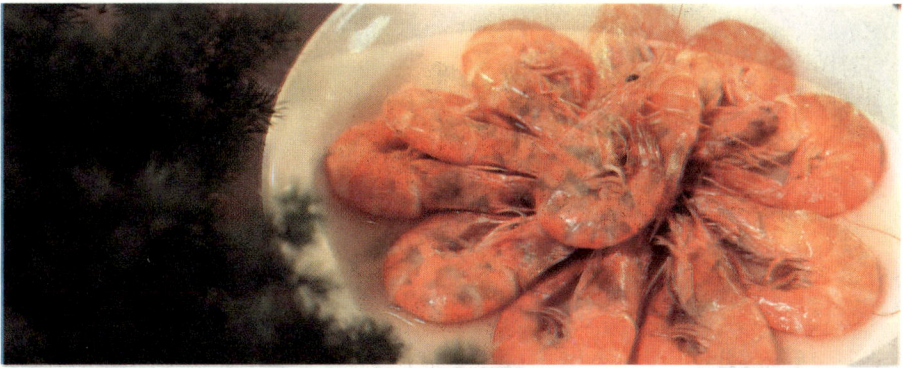

Ingredients
Prawns 600g

Seasonings
Salt 10g
Sichuan peppercorns 5g
1 anise star
Spring onion 10g
Ginger 10g
Vegetable oil 5g
Cooking wine 10g
Water 1,000g

Method
1. Devein the shrimp with a toothpick in the second segment from the tail as shown in the video.
2. Add salt, Sichuan peppercorns, anise star, spring onion, ginger, and water in a wok. Bring it to the boil and simmer on low heat for 1 minute.
3. Plunge prawns in the wok, add the cooking wine and oil. Simmer for 2 minutes, turn heat off, leave the prawns to marinate in the liquid for 5 minutes.
4. Dish out and serve.

Features of the Dish
This dish has a bright colour and umami taste.

Part Three

FISH SLICES WITH PICKLED CUCUMBER

Ingredients
Snakehead fish fillet 200g
Sweet pickled cucumber 30g

Seasonings
Salt 2g
Soy sauce for fish 2g
Sesame oil 15g
Egg 1
Cooking wine 3g
Corn Starch 10g
Sichuan peppercorn oil 10g
Vegetable oil 800g

Method
1. Rinse the fish fillet to clean it well.
2. Slice the fish fillet crosswise at a 45-degree angle, leave in a mixing bowl; chop the pickled cucumber into small pieces, rinse in the cold water and drain well.
3. Mix the fish slices with cooking wine, cornstarch, and a small amount of egg until evenly coated.
4. Pour the vegetable oil in the wok and heat to 120°C, lower the fish slices in, stir to separate them. Blanch briefly in the oil, remove the fish slices from the wok when they turn white. Drain the excessive oil, leave them in a mixing bowl.
5. Blanch the pickled cucumber pieces in boiling water briefly, remove and drain.
6. Add the pickled cucumber pieces to the fish slices in the mixing bowl, season with the salt, soy sauce for fish, sesame oil, and Sichuan peppercorn oil.
7. Mix well and serve immediately.

Features of the Dish
The fish slices are very tender with a rich umami taste.

Notes
In terms of the doneness of the food, all ingredients should be "just cooked" to ensure the tender texture.
To prepare the Sichuan peppercorn oil, pour oil into wok, add Sichuan peppercorns in when the oil is still cold, fry on medium heat until the peppercorns turn black and become aromatic.

STWEET AND SOUR CARP

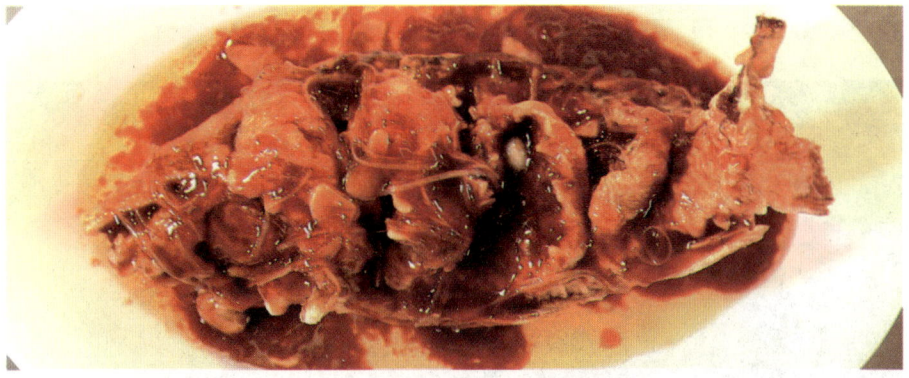

Ingredients
A whole gutted carp fish (about 750g-1,000g)
Shredded Spring onion 10g
Shredded ginger 5g
Sliced garlic 5g

Seasonings
Sugar 300g
Soy sauce 50g
Vinegar 60g
Corn starch 150g
Flour 20g
Clear broth 150g
Salt 5g
Vegetable oil 1.5kg
Tomato sauce 10g
Cooking wine 50g

Method

1. Diagonally score the carp from gill to tail on both sides. Each cut should be at a 45-degree angle until reach the bone.
2. Marinate with cooking wine and salt.
3. Mix water, corn starch (100g), and flour (20g) to make the batter.
4. Pour the vegetable oil into the preheated wok in medium heat.
5. Spread the batter all over the fish, including the inside of the cuts.
6. Hold the tail of the fish with one hand, dip the head into the hot oil, ensure to push the fish away from you as it enters the hot oil, ladle the hot oil over the fish until the head of the fish becomes golden and the batter coating firms up.
7. Leave the whole fish in the hot oil with the tail curled, hold the fish in position with a pair of long-handled chopsticks and a ladle.
8. Remove the fish from the wok when it is golden.
9. Pour the hot oil into the oil drum (oil container).
10. Place the fish in a plate, pat away the excessive oil with a paper towel.
11. Make the starchy seasoning solution by mixing corn starch, water, sugar, salt, soy sauce, tomato sauce, vinegar, cooking wine, and clear broth. Set it aside.
12. Place the wok back on stove, heat up a small amount of vegetable oil on medium heat, stir fry shredded spring onions and ginger, and garlic slices until aromatic.
13. Pour in the starchy seasoning solution, add a bit of oil, stir until the sauce gets thickened.
14. Top the fish with the sizzling sauce and serve.

Features of the Dish

The fish is crispy from the outside and tender and smooth from the inside with a balanced sweet and sour taste. The dish has a dark red colour.

Part Three

Section 2 Meat & Poultry Recipes

KUNG PAO CHICKEN (宫保鸡丁)

Ingredients
Chicken breast 250g
Fried skinless peanuts 30g
Garlic greens 50g
Bamboo shoots 20g
Egg 20g
Vegetable oil 1000g
Corn starch 15g (2x7.5g)

Seasonings
Cooking wine 10g
Soy sauce 6g
Aged vinegar 2g
Broad bean paste 15g
Salt 2g
Sugar 8g
Finely chopped spring onion 5g
Finely chopped ginger 5g
Dried red chili pepper 2g
Red chili oil 40g

Method

1. Cut the chicken breast into 1.5cm x 1.5 cm x 1.5cm pieces.
2. Mix egg, corn starch, salt, and a little water together, pour into the chicken pieces. Combine them well.
3. Cover the bottom of the Wok with water and bring it to the boil - plunge the bamboo shoots into the boiling water, boil for approximately 1 minute and remove.
4. Rinse and reheat the wok to ensure that it is clean and dry. Pour in the vegetable oil and heat it to 120°C, add the chicken breast pieces for an oil-blanching, stir with a ladle to separate them. Add the bamboo shoots in. Remove them from the wok when the chicken pieces are just cooked.

212

5. Place fried skinless peanuts at the bottom of a colander, pour the chicken on top, drain well. Reserve about 2 teaspoons of oil in the wok.
6. Put the wok back on stove, stir-fry the spring onions and ginger until aromatic, add in Pixian broad bean paste, cook until red chili oil is released, then add dried red chilli to the mixture. Continue cooking for a little while so all ingredients are combined well.
7. Stir in oil-blanched chicken dices, season them with the cooking wine, soy sauce, vinegar, salt, and sugar.
8. Add liquid corn starch solution into the wok together with the red chilli oil, stir in garlic greens, continue the stir-frying until the chicken dices are evenly coated.
9. Dish out and serve.

Feature of the Dish
The dish is tender and smooth with a light colour.

Notes
Pixian broad bean paste - or douban jiang - is a fiery hot chilli bean paste, from Pixian in China's Sichuan province. Famed for its chilli heat and umami-rich fermented beans, the Pixian hot bean paste is a key ingredient in many Sichuan dishes. The sauce is commercially available.

Ingredients
Beef 800g
Celery 50g

Ma peppercorns 1g
Dry chili shreds 15g
Chili powder 5g
Salt 1g
Sugar 3g
Oil 1,500g
Corn starch 100g

Seasonings
Sichuan peppercorn 1g
Sichuan peppercorn powder 5g

Method
1. Cut the beef tenderloin to 8cm long, 0.5cm thick shreds.
2. Season the beef with salt 1g and cooking wine 5g, coat evenly with dry corn starch.
3. Cut the celery to 4cm long strips.
4. Heat 1500g of oil up to 210°C, plunge beef in, fry until the meat becomes crispy from the outside and tender from the inside, remove the meant from the hot oil, set aside to drain.
5. Remove the hot oil from the wok and return the wok to the stove.
6. Add a splash of oil in the empty wok, stir fry Pixian broad bean paste. When it is aromatic, add in cooking wine 10g, salt 1g, sugar 3g, and the par-cooked beef, stir fry the beef becomes visibly dry.
7. Add celery, dried chili, Sichuan peppercorns, Ma peppercorns and chili powder, continue to stir fry, remove from heat when celery is just cooked.
8. Dish out and serve.

Feature of the Dish
The dish has a warm dark brown colour. It is a dry with a rich, savoury and aromatic flavour.

Part Three

BEEF TRIPE WITH SHACHA SAUCE

Ingredients
Beef tripe (white) 400g
Coriander 20g

Seasonings
Shacha sauce 20g
Oyster sauce 10g
Sugar 5g
Soy sauce for fish 10g
Sesame oil 15g
Ginger juice 8g

Method
1. Slice the beef tripe into strips and wash well.
2. Boil water (enough to cover the beef tripe) in a wok, immerse the beef tripe in boiling water for about 50 seconds, remove from the wok and plunge into cold water (alternatively ice can be added in), soak for a little while and drain well. Leave the tripe in a mixing bowl.
3. In a separate bowl, mix the Shacha sauce with oyster sauce, ginger juice, sesame oil, soy sauce for fish, and sugar to make the dressing.
4. Add the dressing to the beef tripe and combine well.
5. Plating and serve.

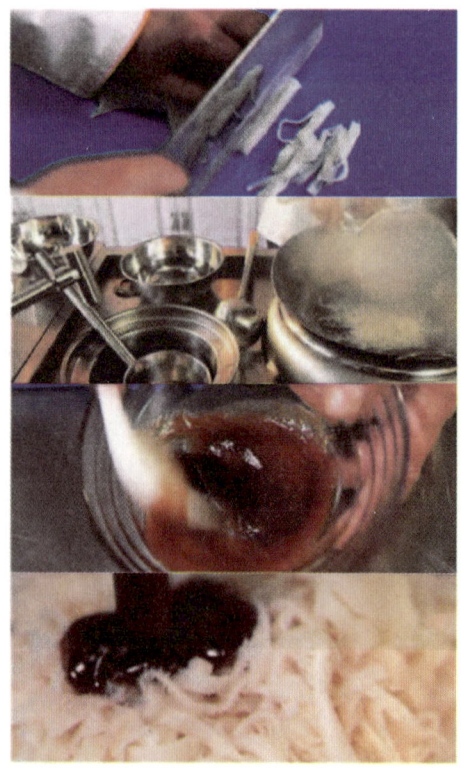

Chinese Culinary Techniques

CHICKEN STRIPS WITH CHAYOTE

Ingredients
Shredded chicken breast 150g
Chayote 50g
Vegetable oil 800g (for frying)

Seasonings
Salt 2g
Sugar 1g
Soy sauce for fish 2g
Sesame oil 15g
Egg 1
Cooking Wine 5g
Corn starch 10g
Sichuan peppercorn oil 10g

Method

1. Rinse and drain the shredded chicken breast, place them in a mixing bowl, add the cooking wine, cornstarch, and a small amount of egg, stir in one direction until they are well combined.
2. Peel and shred chayote.
3. Pour the vegetable oil in the wok and heat it to 120°C, carefully lower the chicken shreds, and cook until the chicken shreds turn white. Remove them from the wok and drain the excessive oil, leave in a mixing bowl.
4. Add the chayote shreds to the mixing bowl, add salt, sugar, soy sauce for fish, sesame oil, Sichuan peppercorn oil, combine well.
5. For plating, square a carrot and slice it until get 7 thin slices, lay them down at the bottom of a serving bowl, place chicken breast and chayote shreds on top and serve.

Part Three

GUAIWEI CHICKEN (怪味白鸡)

Ingredients
A Whole chicken 750g
Spring onions 40g
Ginger 20g

Seasonings
Sichuan peppercorn 16g
Sesame oil 10g
Sesame paste 40g

Hoisin sauce 30g
Oyster sauce 20g
Light soy sauce 10g
Soy sauce for fish 10g
Aged vinegar 30g
Chili oil 40g
Sugar 30g
Salt 2g

Method

1. Remove toes and tail and the wing tips of the chicken, soak it in cold water for 20 minutes to get rid of blood.
2. Place the chicken in a pot of cold water with spring onions and ginger, bring water to the boil on high heat, cook for 1 minute.
3. Turn the heat off, cover the pot with the lid, leave the chicken in hot water to cook for further 25 to 30 minutes.
4. Remove the chicken from the pot, plunge in cold water before immersing in iced water in a mixing bowl*.
5. Place the mixing bowl in the chiller for 15 minutes. Follow the following steps to prepare the Guiwei sauce while wait for the chicken to be chilled. Boil a ladleful of water in the wok.
6. Add the boiling water a small container with 16g Sichuan peppercorns. Leave it to infuse for 15 minutes with the lid on.
7. Remove the Sichuan peppercorns from water, chop them finely.
8. Mix sesame oil and the chopped Sichuan peppercorns in a small bowl, add sesame paste, hoisin sauce, oyster sauce, light soy sauce, soy sauce for fish, aged vinegar, sugar, chili oil, and salt to make the Guaiwei sauce. Take the chicken out of the chiller, and cut in half and chop it Lay the chopped chicken on a plate and top with the Guaiwei sauce serve.

Feature of the Dish
This dish has a spicy, sweet, and sour taste. It is aromatic and provides a unique mouth numbing sensation.

Notes
*If a blast chiller is available to speed the cooling of the ingredients it is recommend to use it.
San huang (or three yellow) chicken is the common choice of ingredient for this dish. It is a popular variety of chicken in Shanghai, China. The three yellows are yellow beak, yellow feet and yellow feathers/skin. The chicken is famous for its tender and juicy texture and a less gamey flavour.

Chinese Culinary Techniques

SLICED COLD PORK WITH GARLIC SAUCE

Ingredients
Pork leg 1,000g
Celery stick 200g

Seasonings
Light soy sauce 30g
Soy sauce for fish 15g
Sugar 15g
Salt 4g
Sesame oil 5g
Minced garlic 10g
Vegetable oil 20g

Method
1. Soak the pork in cold water for 20 minutes to remove blood. Remove from water and roughly halve it.
2. Place the meat in a pot of cold water (enough to cover the meat), bring water to the boil on high heat, remove the white foam.
3. Reduce the heat to low, continue to cook until the pork is 70% done which takes about 25 minutes
4. Remove pork from the pot and in cold water to cool down.
5. Cover the pork with cling film and chill in the fridge for 25 minutes.
6. Pestle minced garlic into garlic paste in a mortar and leave the paste in a small bowl.
7. Heat up 20g of vegetable oil in wok to 120°C.
8. Pour the hot oil onto the garlic paste and let it rest for 20 minutes.
9. Add the light soy sauce, soy sauce for fish, sugar, salt, and sesame oil into the garlic paste, combine well.
10. For the garnish, peel the celery stalk and slice it at a 45° angle, blanch the slices in boiling water. Arrange them into a pattern on a serving plate.
11. Cut the chilled pork into fine slices, roll each slice up to a cylinder, and then place them one by one on the serving plate.
12. Top with the garlic sauce and serve.

Feature of the Dish
The dish is tender and umami with a strong garlic aroma. It is fatty but not greasy.

Part Three

RED LU AROMATIC SQUAB

Ingredients

Squabs 1,000g
Celery 150g
Coriander 45g
Carrots 200g
Onions 30g
Ginger 30g
Green peppers 100g
Dried scallop 80g
Dried shrimps 40g
Dry cured ham 40g
Water 4,000g

Seasonings

Light soy sauce 90g
Dark soy sauce 35g
Maggi seasoning soy sauce 15g
Fish sauce 30g
Lea & Perrins sauce 15g
Soy sauce for fish 15g
Rock sugar 40g
Cooking wine 60g
Black peppercorns 10g
Salt 10g
Sha Ren (Fructus Amomi) 8g
White Cardamom 10g
White Angelica Root 15g
Cao Guo (Tsao-Ko Cardamom) 5g
Fennel Seed 8g
Luo Han Guo 15g
Sichuan peppercorns 6g
Star Anise 8g
Bay leaf 3g
Lemongrass 5g

Method

1. Remove the feet and leave the squabs in cold water to soak for 20 minutes.
2. Bring water (enough to cover the birds) to a boil in wok, plunge the squabs in, blanch briefly and remove, leave the birds to cool in cold water.
3. To make red broth:
    a. Add the black peppercorns 10g, sha ren (fructus amomi) 8g, white cardamom 10g, white angelica root 15g, cao guo (tsao-ko cardamom) 5g, fennel seed 8g, Luo han guo 15g, Sichuan peppercorns 6g, star anise 8g, bay leaf 3g, and lemongrass 5g in a muslin food bag. Tie the bag up and leave it in the pot with 4,000g water.

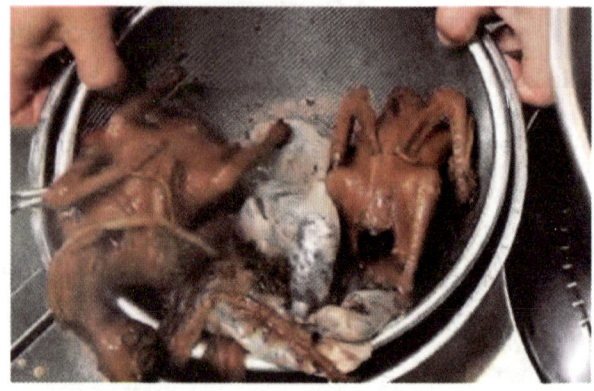

- b. Add dried scallop 80g, dried shrimps 40g, and dry-cured ham 40g into the second muslin bag. Tie it up and plunge into the pot.
- c. Drop in Celery 150g, coriander 45g, carrots 200g, onions 30g, ginger 30g, and green peppers 100g to the pot.
- d. Season with light soy sauce 90g, dark soy sauce 35g, maggi seasoning soy sauce 15g, fish sauce 30g, Lea & Perrins sauce15g, soy sauce for fish 15g, rock sugar 40g, and cooking wine 60g.
- e. Bring water to the boil and cook on high heat for 20 minutes.
4. Plunge the squabs in the red broth and simmer on low heat for 40 minutes.
5. Remove squabs from the red broth.
6. Brush a layer of oil onto the squabs.
7. Cut the squab into smaller pieces and serve.

Feature of the Dish

The dish is red brownish in colour. The squabs are tender and moist with a strong aroma.

Notes

Aromatic herbs and vegetables and dried seafood products are the best ingredients to make light yet umami rich broth.

Part Three

JIANG BEEF

Ingredients

Beef shank 2,000g
Water 4,000g

Seasonings

Soy bean paste 80g
Donggu soy sauce 70g
Dark soy sauce 30g
Oyster sauce 60g
Cooking wine 60g
Rock sugar 40g
Salt 20g
Chicken stock powder 20g
Dried chilli 6g

Spring onion 160g
Ginger 50g
Celery 100g
Beef sauce 40g
Start anise 16g
Cinnamon 15g
Clove 2g
Cao guo (Tsao-ko cardemom) 5g
Fennel seeds 8g
Luo han guo 15g
Sichuan peppercorns 6g
Bay leaf 3g
Oil 15g

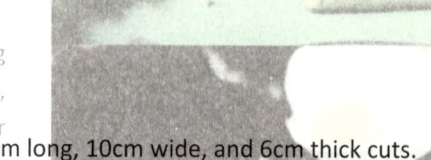

Method

1. Cut the beef shank crosswise into approximately 15cm long, 10cm wide, and 6cm thick cuts.
2. Leave the meat in cold water to soak for 4 hours.
3. Remove from water, drain well. Piecre the beef cuts in several places with bamboo sticks. It will help the marinade sink deeper into the beef.
4. In a mixing bowl, add the dark soy sauce 30g, cooking wine 30g, salt 10g, spring onion 80g, and ginger 25g, place the beef cuts in.
5. Cover the mixing bowl and leave it to marinate in the chiller for 24 hours.
6. Once marination is done, take the beef cuts out and leave them in cold water in a wok, bring water to the boil, cook on high heat for 15 minutes. Remove white foam that floats on the surface.
7. Take beef out of the wok, cool in cold water or blast chiller if it is available.
8. Remove the beef from cold water/blast chiller, drain well, cut off excess fat, and set aside.

220

9. Pour a small acount of oil in the wok, stir fry spring onions 80g and giner 25g until aromatic. Add the rest of the ingredients and seasongs in.
10. Add 4,000g cold water into the wok, bring it to the boil, drop in beef cuts, cook for 2 hours on low heat.
11. Reduce the broth by using medium heat in the last 5 minutes to create a enhanced taste and colour. If it is done properly, it should not have too much liquid left in the wok.
12. Take the beef pieces out, cool them down on a plate, brush a layer of oil on the meat.
13. Slice the meant and serve.

Feature of the Dish
The dish is dark red brown in colour, tender, savoury and aromatic.

Notes
To achieve the correct texture, the marinated meat should be par-cooked following the process of placing the meat in cold water in wok, bringing water to the boil, boiling the meat on high heat for 15 minutes, removing the meat from the wok, cooling the meat in cold water or blast chiller.

Ingredients	Seasonings
Chicken breast 300g	Vegetable oil 1,500g
Bamboo shoots dices 50g	Sweet flour paste 50g
	Sugar 30g
	Salt 1g
	Soy sauce 30g
	Cooking wine 20g
	Cornflour 25g
	Egg 25g
	Sesame oil 15g
	Ginger juice 10g

Method
1. Dice the chicken breast and bamboo shoots.
2. In a bowl, add salt, egg, corn starch, use your hand to mix them well.
3. Pre-heat wok on medium to low heat, pour the vegetable oil in, bring it to 150°C, carefully lower the chicken dices in with the help of a pair of chopsticks, quickly stir them in the hot oil to separate them, add the bamboo shoots in, blanch briefly, remove the wok from heat.
4. Pour the chicken and bamboo shoots with the hot oil in a colander, let them to drain.
5. Place the empty wok back on heat, add a small amount of oil, stir fry sweet flour paste, ginger juice, cooking wine, sugar, and soy sauce until thickened.
6. Add chicken and bamboo shoots dices in, sprinkle sesame oil on, rapidly stir the ingredients and toss the wok.
7. Dish out and serve.

Feature of the Dish
This dish has a good balance of slaty and sweet flavour with a strong fermented flour paste aroma.

Part Three

ZAO PIG TROTTER

Ingredients
Pork trotter 1,000g

Seasonings
Zao sauce 100g
Cooking wine 40g
Salt 12g
Spring onion 20g
Ginger 20g
Sichuan peppercorns 5g
Star aniseed 5g
Cinnamon 2g

Method
1. Halve the trotter lengthwise, put them in pot with cold water.
2. Bring water to the boil, simmer for 50 minutes on low heat.
3. Take the trotter pieces out and place on the chopping board.
4. Remove the big bone, rub salt in, leave the meat aside.
5. Use half of the broth in the pot, add spring onions, star aniseed, cinnamon, Sichuan peppercorn, salt, cooking wine, and zao sauce.
6. Bring broth to the boil.
7. Remove liquid from the wok and filter into a mixing bowl.
8. Soak trotter in the liquid.
9. Store in the fridge for a day.
10. Remove from the fridge.
11. Roughly cut the meat into smaller pieces.
12. Serve.

Feature of the Dish
This dish has a unique fermented rice wine flavour and a tender texture.

Chinese Culinary Techniques

FRIED LAMB SKEWERS

Ingredients
Lamb tenderloin 400g

Seasonings
White pepper powder 2g
Salt 2g
Minced onion (white) 10g
Chili powder 10g
Cumin powder 10g
Sichuan pepper powder 2g
Vegetable oil 1,500g

Method
1. Infuse Sichuan peppercorns in 5ml boiling water for 30 minutes to make Sichuan peppercorn infused water.
2. Cut lamb into 2.5cm x 2.5cm dices, season with pepper 2g, salt 5g, chili powder 10g, and minced onion 10g, leave the meat aside to marinate for 20 minutes.
3. Skewer the lamb dices.
4. Heat wok up, pour vegetable oil in, heat the oil to 80°C ~ 120°C on medium heat, plunge lamb skewers in. Fry until golden.
5. Remove the skewers from the hot oil, sprinkle cumin powder 6g and chili powder 2g on top.
6. Mix the remaining seasonings together and serve it as a side dip with the lamb skewers.

Feature of the Dish
The lamb skewers are chewy and aromatic.

223

Part Three

AROMATIC CRISPY CHICKEN

Ingredients

A whole chicken 1,500g – 2,000g

Seasonings

Ginger 10g
Spring onions 10g
Sichuan peppercorns 5g
Corn starch 30g
White cardamom 1 pod
Bay leave 1g
Funnel seeds 0.5g
Cassia bark (Chinese cinnamon) 5g
Salt 12g
Cooking wine 25g
Sichuan pepper salt 5g (for dipping)
Vegetable oil 2,000g

Method

1. Remove chicken feet and tip of the wings, butterfly the chicken, and rinse it in water.
2. Place the chicken in a mixing bowl, season it with salt 12g, spring onions 10g, ginger 10g, cooking wine 25g, Sichuan peppercorns 5g, 1 white cardamom pod, bay leaves 1g, funnel seeds 0.5g, and Cassia bark 5g. Message the bird well. Leave it to marinate for 30minutes.
3. Steam the bird for about 45 minutes or until it is tender (see Tips).
4. Remove the spices and herbs, place the chicken on a clean plate, dust some dry flour on it, press gently to make sure that the flour stays on.
5. Heat up 2,000g vegetable oil in wok to 210°C, carefully lower the chicken in.
6. Fry until chicken skin becomes crispy and turns to golden.
7. Remove the chicken from the hot oil, drain it well.
8. Place the chicken on the chopping board with its breast side facing up, chop it into bit-size pieces.
9. Serve with Sichuan pepper salt as the side dip.

Tips

To check if the chicken is steam cooked, use a chopstick to poke the thigh or breast. It is done if chopstick can easily go through.

Feature of the Dish

This dish has a savoury and umami taste with a strong aroma. The bird is tender on the inside and crispy on the outside with an lovely golden colour.

SHREDDED PORK WITH SWEET PEPPERS

Ingredients
Pork Loin 200g
Green and red sweet peppers 300g

Seasonings
Sesame oil 1g
Cooking wine 5g
Soy sauce 5g
Salt 3g
Corn starch 4g
Vegetable oil 15g
Spring onions shreds 5g
Ginger shreds 5g

Method
1. Cut pork loin into fine shreds.
2. Season the pork shreds with salt 1g, corn starch, and 1 tablespoon of cold water, leave the meat to marinate for 15 minutes.
3. Shred the green and red sweet peppers into the same sized of the pork loin.
4. Pre-heat the work, add a splash of vegetable oil, stir fry the peppers briefly and remove them from the wok.
5. Place the wok back on stove, pour a small amount of vegetable oil in, stir fry the spring onions and ginger until aromatic, and then add the pork in, continue to stir fry until the pork turns white.
6. Return the par cooked sweet peppers to the wok, season with cooking wine, soy sauce, salt, and sesame oil. Rapidly stir the ingredients with the ladle while keep tossing the wok.
7. Dish out and serve.

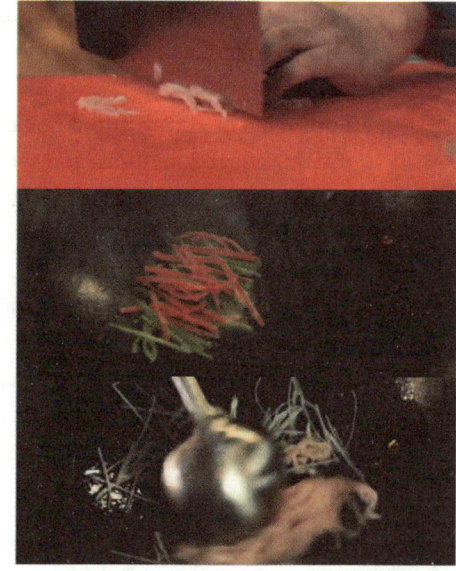

Features of the Dish
With a combined flavour of umami and savoury, the pork shreds are smooth and tender while the sweet pepper are crunchy and refreshing.

Part Three

TWICE COOKED PORK

Ingredients
Boneless pork leg with skin 400g
Garlic greens 75g

Seasonings
Pixian broad bean paste 50g
Sweet flour paste (or fermented flour paste) 10g
White sugar 3g
Cooking wine 5g
Soy sauce 10g
Salt 2g
Vegetable oil 40g
Spring onion 5g
Ginger 5g
Anise stars 2
Garlic greens 5g
Leek 50g (roughly chopped)

Method
1. Wash and score the surface of the meat all over with a cleaver.
2. Cover the pork in cold water and add the leek, ginger, and anise stars.
3. Put the lid on, bring it to a boil on medium heat, simmer for about 8 minutes, and set aside to cool.
4. Once cooled down, slice meat into 5cm long, 4cm wide, 0.2cm thick slices.
5. Cut garlic greens into segments.
6. Heat wok up on medium heat, add a small amount of oil, stir fry pork slices, add in cooking wine 5g, Pixian broad bean paste 50g, sweet flour paste 10g, sugar 3g, soy sauce 10g, salt 2g, and garlic greens.
7. Stir occasionally while adding in seasonings, to avoid burning, cook unit it is aromatic.
8. Dish out and serve.

Feature of the Dish
Notes
Depending on the quality of meat, the cooking time may variy. Use a chopstick to poke the meat to check if the skin is tender and the meat is cooked from the inside.

Chinese Culinary Techniques

SALTED PORK SHANK

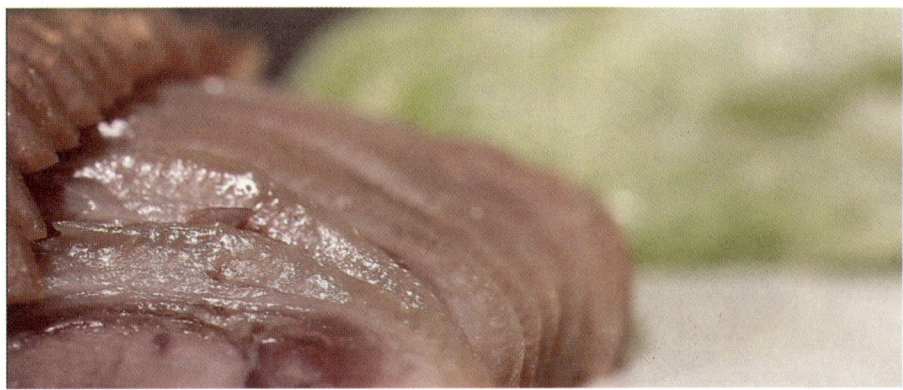

Ingredients
Pork shank 1,500g
Lettuce 90g

Seasonings
Yellow cooking wine 45g
Salt 30g
Spring onion 50g
Ginger 40g

Sichuan peppercorns 6g
Fennel 5g
Bay leaves 2g
Black cardamom 5g
Amomi fructus rotundus 8g
Minced ginger 20g
Aged vinegar 30g
Light soy sauce 15g
Sesame oil 10g

Method

1. Soak the pork shank in cold water for 30 minutes.
2. Debone.
3. Rub 30g of onto the meat, leave it to marinate for 2 hours.
4. Place the meat flat on the chopping board, roll it tight from one end, wrap it with cheese cloth, and finally use cooking string to tie it up.
5. Place the meat roll in cold water in a wok. Bring it to the boil and cook for 10 minutes.

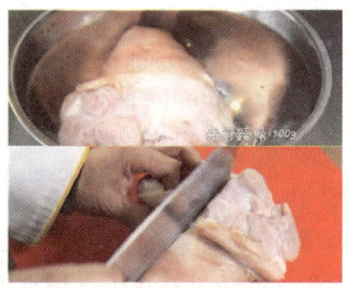

6. Remove the shank from the boiling water and soak it in a pot with cold water, add in spring onions 50g, ginger 40g, Sichuan peppercorn 6g, fennel 5g, bay leaves 2g, black cardamon 5g, amomi fructus rotundus 8g, and cooking wine 35g. Bring it to the boil, boil the meat on midium heat for 25 minutes.
7. Remove the pork from pot, leave it aside to cool down.
8. Once it cools down, wrap with cling film and place it in the chill blaster for 20 minutes.
9. Slice the shank roll into thin pieces and lay them on the vegetable bed made of iceberg lettuce.
10. Prepare the dipping sauce by mixing minced ginger 20g, aged vinegar 30, light soy sauce 15g, and sesame oil 10g.
11. Serve.

Feature of the Dish
Salted port shank is savoury and aromatic, and not greasy.

Notes
Make sure the meat is firmly rolled up and tightly tied up. This will make a nice-looking meat roll when it is cooked.

Part Three

MALA CHICKEN (麻辣白鸡)

Ingredients
A small Chicken 750g

Seasonings
Spring onions 44g
Ginger 20g
Sichuan peppercorns 16g
Sesame oil 10g
Light soy sauce 34g
Soy sauce for fish 15g
Old/aged vinegar 22g
Sugar 8g
Spicy oil 20g
Salt 4g

Method
1. Remove toes and tail and the wing tips of the chicken, soak it in cold water for 20 minutes to get rid of blood.
2. Place the chicken in a pot of cold water with spring onions and ginger, bring water to the boil on high heat, cook for 1 minute.
3. Turn the heat off, cover the pot with the lid, leave the chicken in hot water to cook for further 25 to 30 minutes.
4. Remove the chicken from the pot, plunge in cold water before immersing in iced water in a mixing bowl*.
5. Place the mixing bowl in the chiller for 15 minutes.

Follow the following steps to prepare the Mala sauce while wait for the chicken to be chilled.

6. Place Sichuan peppercorns 16g in a small bowl, add enough boiling water to cover them, and leave to infuse for 15 minutes with the lid on.
7. Remove the peppercorns from water, chop them finely.
8. Mix sesame oil and the chopped peppercorns in a small bowl, add in the remaining seasonings.

To serve
9. Take the chicken out of the chiller and cut it into bite-size pieces.
10. Lay the chopped chicken on a plate and top with the Mala sauce.
11. Serve.

Feature of the Dish
This chicken dish has a smooth and tender texture, it is spicy with a mouth-numbing sensation which is very appetizing.

Notes
*If a blast chiller is available to speed the cooling of the ingredients, it is recommended use it.
Similar to Guiwei chicken, the san huang (or three yellow) chicken is a popular choice of ingredient for this dish.

DRY STIR-FRIED BEEF

Ingredients
Beef 800g
Celery 50g

Seasonings
Sichuan peppercorn 1g
Sichuan peppercorn powder 5g
Ma peppercorns 1g
Dry chili shreds 15g
Chili powder 5g
Salt 1g
Sugar 3g
Oil 1,500g
Corn starch 100g

Method
1. Cut the beef tenderloin to 8cm long, 0.5cm thick shreds.
2. Season the beef with salt 1g and cooking wine 5g, coat evenly with dry corn starch.
3. Cut the celery to 4cm long strips.
4. Heat 1500g of oil up to 210°C, plunge beef in, fry until the meat becomes crispy from the outside and tender from the inside, remove the meant from the hot oil, set aside to drain.
5. Remove the hot oil from the wok and return the wok to the stove.
6. Add a splash of oil in the empty wok, stir fry Pixian broad bean paste. When it is aromatic, add in cooking wine 10g, salt 1g, sugar 3g, and the par-cooked beef, stir fry the beef becomes visibly dry.
7. Add celery, dried chili, Sichuan peppercorns, Ma peppercorns and chili powder, continue to stir fry, remove from heat when celery is just cooked.
8. Dish out and serve.

Feature of the Dish
The dish has a warm dark brown colour. It is a dry with a rich, savoury and aromatic flavour.

Part Three

MOIST STIR-FRIED PORK LOIN

Ingredients
Pork tenderloin 200g
Bamboo shoots 40g

Seasonings
Salt 5g
Cooking wine 12g

Egg 20g
Broth 25g
Corn starch 10g
Vegetable oil 750g
Spring onion 3g
Ginger 3g

Method
1. Rinse and cut pork lion into fine shreds (5cm long, 0.2cm wide).
2. Cut bamboo shoots into shreds (approx. 4cm long, 0.2cm thick).
3. Marinate the pork with a small amount of egg white, corn starch, and salt 2g in a mixing bowl.
4. Shred the spring onions and ginger.
5. Make seasoning sauce by mixing salt 3g, cooking wine, and corn starch in a bowl.
6. Blanch the bamboo shoot in oil or water.
7. Heat 750g of vegetable oil up in wok to 120°C, add the pork in, constantly stir the pork for even cooking.
8. Fry until the meat turns white, remove it from the hot oil, set aside to drain.
9. Pour 50g of vegetable oil in the empty wok, heat up to 210°C, plunge spring onions and ginger in, stir fry until aromatic, add in bamboo shoots, pork and the seasoning sauce, continue with stir frying, add a splash of oil on top.
10. Dish out and serve.

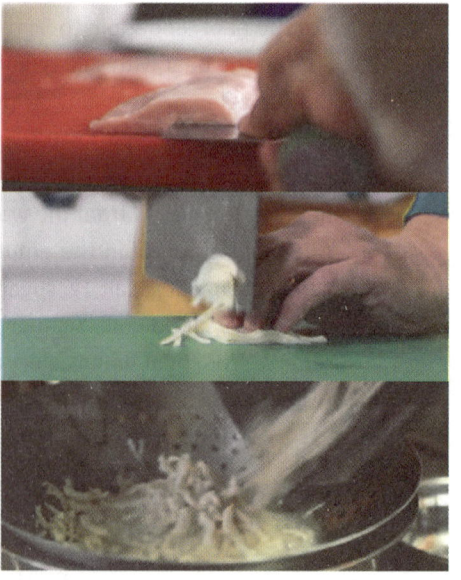

Feature of the Dish
The pork is tender and smooth with a light colour.

BEEF WITH OYSTER SAUCE

Ingredients
Beef 400g

Seasonings
Vegetable oil 1,500g
White pepper powder 3g
Salt 3g
Onion 15g
Oyster sauce 15g
Cooking wine 10g
Sugar 5g
Light soy sauce 5g
Dark soy sauce 5g
Corn starch 20g
Egg 20g
Red sweet pepper 4g
Green sweet pepper 4g
Finely chopped ginger 3g

Method
1. Slice the beef, soak in water to remove the blood, drain well.
2. Season the beef with light soy sauce and salt, then mix them with egg and corn starch.
3. Slice the onion, red and green sweet peppers.
4. Pour vegetable oil into the pre-heat wok, bring it to 120°C, carefully stir in beef slices, fry them in the hot oil until nearly cooked, add the green and red sweet pepper in, cook briefly, empty the wok onto a colander, leave the beef and sweet peppers in the colander to drain.
5. Place the empty wok back to stove, stir fry the minced ginger until aromatic with a small amount of oil, mix in oyster sauce, white pepper powder, add cooking wine, light soy sauce, dark soy sauce, salt, sugar, and a small amount of water, continue with the stir-frying.
6. Thicken the dish with corn starch solution, add a splash of oil.
7. Dish out and serve.

Feature of the Dish
The beef is smooth and tender with a savoury and umami taste.

Part Three

YUXIANG SHREDDED CHICKEN

Ingredients

Chicken breast 250g
Green and red sweet pepper 30g
Bamboo shoots 10g
Rehydrated dried shiitake mushrooms 3

Seasonings

Pixian broad bean paste 10g
Pickled red chili 10g
Sugar 15g
Cooking wine 10g

Soy sauce 5g
Vinegar 8g
Salt 2g
Corn starch 15g
Egg 5g
Shredded spring onions 5g
Shredded ginger 5g
Sliced Garlic 5g
Red chili oil 5g
Sesame oil 5g
Vegetable oil 1,000g

Method

1. Shred the chicken breast, green and red sweet pepper, shiitake mushrooms, and bamboo shoots.
2. Mix the chicken breast with egg 5g, corn starch 10g, and salt 1g.
3. Blanch green and red sweet peppers, shiitake mushrooms, and bamboo shoots in boiling water briefly, remove and drain well.
4. Pour 1,000g of oil into the pre-heated wok, bring the oil to 90-120°C, lower in the chicken breast. Use a pair of long handled chopsticks to separate the chicken pieces by stirring. Plunge the green and red sweet peppers, shiitake mushrooms, and bamboo shoots in for a quick oil blanching when the chicken breast is nearly cooked. Pour the hot oil onto a colander to leave all the ingredients in the colander to drain.
5. Stir fry shredded ginger, spring onions, and garlic slices with a small amount of oil in the wok until aromatic, add the pickled red chili, broad bean pastes to the mixture, cook until the red chili oil emits.
6. Add in all the oil blanched ingredients, season them with cooking wine, soy sauce, vinegar, salt, and sugar. Splash a small amount of water to the ingredients, stir them constantly with a ladle and toss the wok every now and then.
7. Finally, add in corn starch solution to thicken the dish, add a splash of red chili oil and sesame oil as the final touch.
8. Dish out and serve.

Feature of the Dish

The dish looks moist and colourful. It has a complex flavour called "Yuxiang "which is salty, sweet, spicy, aromatic, and sour.

GULAO PORK WITH PINEAPPLE

Ingredients
Pork leg 200g
Pineapple 50g
Green sweet pepper 15g
Red sweet pepper 15g

Seasonings
Baking powder 4g
Flour 12g
Corn starch 40g
Cheese powder 3g
Salt 2g
Sugar 120g
Spring onion and ginger infused water 8g
Cooking wine 15g
White vinegar 100g
Tomato sauce 20g
Water 30g
Vegetable oil 1,000g

Method
1. Slice the pork into dices, mix with cooking wine, spring onions and ginger infused water, and salt in a mixing bowl. Leave the meat aside to marinate.
2. In a separate mixing bowl, mix baking powder, cheese powder, corn starch, flour, water, and cooking wine together until form a thick batter.
3. In another bowl, mix tomato sauce with vinegar, sugar, salt, cooking wine, corn starch solution, and water to make the starchy seasoning sauce.
4. Cut the pineapple to dices.
5. Cut the red and green peppers obliquely into small pieces.
6. Bring the oil to 130°C, dip the pork dices in the batter to have them coated evenly. Carefully lower them one by one into the hot oil.
7. Fry the pork until crust forms up around each piece. Remove them from the hot oil, set aside to drain.

8. Bring the oil to 210°C -240°C, plunge the pork dices back in, fry briefly, add in pineapple, red and green peppers, remove the wok from heat, pour the ingredients onto a colander to drain.
9. Place the empty wok back on stove, stir fry the starchy seasoning sauce rapidly, add in a little bit oil, and plunge all the ingredients into the work.
10. Toss the wok a few times to make sure the ingredients and the sauce are mixed well.
11. Dish out and serve.

Feature of the Dish
The pork is crispy from the outside and tender from the inside with a sweet and sour flavour and an orange colour.

JIAO LIU PORK LOIN

Ingredients
Pork tenderloin 250g

Seasonings
Cooking wine 15g
Soy sauce 10g
Vinegar 10g
Salt 3g
Shredded spring onion 5g
Shredded ginger 5g
Garlic slices 5g
Spring onions and ginger infused water 8g
Egg 10g
Corn starch 15g
Flour 5g
Vegetable oil 1,000g
Sesame oil 10g

Method
1. Slice the spring onions and ginger into fine shreds and garlic into slices.
2. Slice the pork tenderloin, marinate them in salt, cooking wine, spring onions and ginger infused water.
3. Mix egg, corn starch, and flour to make the batter.
4. In a bowl, make the starchy seasoning sauce by mixing cooking wine, soy sauce, vinegar, salt, spring onions, ginger, garlic, and a little bit water.
5. In a pre-heated wok, bring the 1,000g of oil to 120°C, dip the pork slices into the batter to have them coated evenly. Carefully plunge the pork slices one by one into the hot oil.
6. Fry until crust forms up around each piece, remove them from the hot oil and set aside to drain.
7. Bring the oil to 210°C -240°C, plunge the pork slices back in, remove when they turn light golden.
8. Place the empty wok back on stove, stir fry pork slices rapidly with the starchy seasoning sauce. Sprinkle in a splash of oil.
9. Quickly toss the wok a few times to make sure the pork slices and the sauce are combined well.
10. Dish out and serve.

Feature of the Dish
The pork is golden in colour, tender from the inside and crispy from the outside with a salty, umami, and aromatic and sour taste.

Chinese Culinary Techniques

MOIST LIU PORK LION

Ingredients
Pork tenderloin 250g
Black ear fungus (wood ear fungus) 10g
Bamboo shoots 10g

Seasonings
Vegetable oil 500g
Salt 4g
Sugar 5g
Vinegar 20g
Cooking wine 5g
Spring onion (2.5g) and ginger (2.5g) infused water
Corn starch with water 20g
Egg white 150g

Method
1. Shred the pork tenderloin, leave them in a mixing bowl.
2. Slice the bamboo shoots, leave them aside.
3. In the mixing bowl, season the shredded pork with salt 4g, corn starch 8g, and egg white.
4. Blanch bamboo shoots and black ear fungus briefly in boiling water, remove from wok to drain.
5. Preheat the wok, pour 500g of vegetable oil in, heat it up to 100°C ~ 120°C, lower the prepared pork shreds into the hot oil, rapidly stir the pork shreds to separate them from each other, remove from the hot oil when pork shreds turn white.
6. Carefully empty the wok of oil.
7. Place the empty wok back on stove, add in the spring onions and ginger infused water, cooking wine, salt, clear broth, and corn starch solution. Bring the sauce to the boil and add black ear fungus, bamboo shoots, pork shreds in, add a bit of oil while stirring.
8. Dish out and serve.

Feature of the Dish
The pork is white in colour, tender, smooth, clean and aromatic, with a balanced savoury and umami flavour.

Part Three

OIL BAO DICED CHICHEN

Ingredients
Chicken breast 250g
Cucumbers 100g

Seasonings
Salt 5g
Cooking wine 10g
Vinegar 2g

Minced spring onion 5g
Sliced garlics 10g
Corn starch 20g
Corn starch solution 5g
Vegetable oil 1,200g
Egg white 20g
A small a mount of clear broth or water

Method
1. Dice the chicken breast into 1.5cm x 1.5cm cubes.
2. Remove the seeds and dice the cucumber into 1.5cm x 1.5cm cubes.
3. Mix the chicken dices well with salt, egg white 10g, and corn starch 15g.
4. Prepare the seasoning sauce in a bowl by mixing cooking wine, salt 3g, vinegar, minced spring onions, garlic slices, corn starch solution 5g, clear broth or water.
5. Heat the wok up on medium to low heat, pour oil, bring it to 120°C, carefully scatter the chicken breast in with the help of a pair of chopsticks, quickly move the dices around in the hot oil to separate them, remove the wok from heat.
6. Pour the chicken breast with hot oil on to the cucumber dices in a colander so that the cucumber is "washed" with the hot oil, leave them to drain.
7. Plunge the chicken and cucumber dices back in wok before placing the wok back on heat, add the beaten egg white 10g, stir rapidly with a ladle, add a small amount of oil, quickly toss the wok while stirring with the ladle.
8. Dish out and serve.

Feature of the Dish
The dish is crunchy, tender, refreshing, savoury and umami with a garlic aroma.

Chinese Culinary Techniques

BAO CHICKEN WITH SWEET FLOUR PASTE

Ingredients
Chicken breast 300g
Bamboo shoots dices 50g

Seasonings
Vegetable oil 1,500g
Sweet flour paste 50g
Sugar 50g
Salt 1g
Soy sauce 30g
Cooking wine 20g
Cornflour 25g
Egg 25g
Sesame oil 15g
Ginger juice 10g

Method
1. Dice the chicken breast and bamboo shoots.
2. In a bowl, add salt, egg, corn starch, use your hand to mix them well.
3. Pre-heat wok on medium to low heat, pour the vegetable oil in, bring it to 150°C, carefully lower the chicken dices in with the help of a pair of chopsticks, quickly stir them in the hot oil to separate them, add the bamboo shoots in, blanch briefly, remove the wok from heat.
4. Pour the chicken and bamboo shoots with the hot oil in a colander, let them to drain.
5. Place the empty wok back on heat, add a small amount of oil, stir fry sweet flour paste, ginger juice, cooking wine, sugar, and soy sauce until thickened.
6. Add chicken and bamboo shoots dices in, sprinkle sesame oil on, rapidly stir the ingredients and toss the wok.
7. Dish out and serve.

Feature of the Dish
This dish has a good balance of slaty and sweet flavour with a strong fermented flour paste aroma.

MENN CHICKEN WITH CHESTNUTS

Ingredients
Chicken legs 2
Shelled chestnuts 40g

Seasonings
Cooking wine 30g
Salt 6g
Sugar 15g

Spring onion 5g
Ginger 5g
Soy sauce 50g
Vegetable oil 1,500g
Chicken stock 350g
Corn starch 20g
Anise oil 10g
Sichuan peppercorn oil 10g

Method

1. Debone the chicken leg, cut them into cubes (4cm), mix well with cooking wine 10g, cornflour, and salt 1g. Set them aside to marinate. Finely chop the ginger and spring onions.
2. Pre-heat the wok on medium heat, pour oil in, bring it to 150°C, fry the chestnuts until it is just cooked, remove them from the hot oil with a colander, let them to drain.
3. Bring the oil temperature to 210°C, blanch the chicken leg pieces, remove them from the hot oil with the colander, leave them to drain.
4. Place the empty wok back on heat, add in anise oil, stir fry finely chopped spring onions and ginger until aromatic, plunge the chicken and chestnuts in, and then add cooking wine, soy sauce, sugar, salt, and chicken stock, bring it to the boil, remove scum if any, put a lid on, cook on low heat until all the ingredients become tender, and the flavour of the broth set in the chicken and chestnuts.
4. Thicken the broth on high heat until it is visibly disappeared, sprinkle the Sichuan peppercorn oil. In the whole cooking, it is important to toss the wok constantly to avoid food gets stuck on the wok.
5. Dish out and serve.

Feature of the Dish
This dish is glossy light brown in colour, the chicken is soft and tender, and the chestnuts is sweet and aromatic.

Chinese Culinary Techniques

DUN CHICKEN

Ingredients
A gutted whole chicken (about 1kg)

Seasonings
Salt 5g
Cooking wine 10g
Spring onion finely chopped 10g
Ginger slices 15g
White pepper powder 1g
Water 500g

Method
1. Remove the beak and nails of the chicken.
2. Place the chicken in a ceramic casserole pot, add in all seasonings except for the salt, add 500g of water, seal the pot with cling film.
3. Leave the pot in a steamer to steam with the lid on for 3 hours on medium heat or until the chicken is soft and tender.
4. Add in the salt and continue with the steaming for another 10 minutes, remove from the steamer.
5. Serve in small soup bowls.

Feature of the Dish
The soup is clear, fresh, and umami while the chicken is tender and aromatic.

Part Three

DUN MEATBALLS

Ingredients
Pork belly (40% fat) 800g
4 Pak choi hearts and 4 leaves 250g

Seasonings
Cooking wine 100g
Salt 10g

Spring onion and ginger infused water 300g
Cornstarch 25g
Vegetable oil 50g
Supreme broth (premade bone soup) 500g

Method

1. Cut the pork into pomegranate seed size, mix them with cooking wine, spring onions and ginger infused water, salt, and corn starch. Use your hand to blend them well.
2. Cross-score on the head of the pak choi, trim the pak choi leaves to tidy them up.
3. Set wok on medium heat, add in 40g of vegetable oil, quickly stir fry the pak choi heart unit it becomes bright green.
4. Add salt and pork broth, bring it to the boil, turn the heat off, remove the vegetable.
5. Grease a saucepan with 10g of vegetable, lay the 4 pak choi hearts at the bottom to form a cross.
6. Carefully pour the supreme broth in.
7. Sperate the pork mince into four equal portions, make them into four large -smooth meat balls, lay them one by one in the space between the pak choi hearts, cover them with the leaves.
8. Simmer for 2 hours on minimum heat with the lid on.
9. Dish out and serve.

Feature of the Dish
The meatballs have a light and fatty texture but not greasy. The soup is clear and umami.

Chinese Culinary Techniques

STEAMED PORK WIT RICE AND LOTUS LEAVES

Ingredients
Pork belly 600g
Short grain rice 100g
Fresh lotus leaves (rehydrated from the dried lotus leaves) 2

Seasonings
Shredded spring onion 30g
Shredded ginger 30g
Dried sand ginger 0.5g
Star anise 0.5g
Clove 0.5g
Chinese cinnamon (cassia bark) 1g
Cooked fermented flour paste (stir-fried) 75g
Sugar 15g
Cooking wine 40g
Vegetable oil 75g
Soy sauce 10g

Method

1. Dry stir fry (on oil) the rice with sand ginger, star anise, clove, and cinnamon on low heat until the rice turns slightly brown, remove it from the wok.
2. Pick the spices out and grind them into small grain size in a grinder, add the rice in, continue with the grinding until the mixture becomes powdery.
3. Score the entire surface of the meat so that flavour could set in easier, rinse well, cut it into long thin slices (0.5cm thick).
4. Marinate the pork slices with stir-fried fermented flour paste, soy sauce, sugar, cooking wine, shredded spring onions and ginger for an hour.
5. Blanch the lotus leaves in hot water and spread it in a steamer basket.
6. Dip a slice of pork in the rice powder to make sure it is evenly coated, place it on the lotus leaves in the steamer basket.
7. Repeat this step with all the pork slices. Now, wrap the leave up, cover the steamer basket with the lid.
8. Bring water in a pot to the boil on medium heat, place the steamer basket in, cover the pot with the lid, steam for 2 hours.
9. Serve.

Feature of the Dish
The pork is tender, gummy, aromatic, and non-greasy.

CUAN MEATBALLS

Ingredients
Pork mince 250g
Rehydrated black ear fungus 20g
Finely chopped coriander 10g

Seasonings
Cooking wine 15g
Salt 3g
Egg white 1
Corn starch 10g
Supreme broth 750g
Vegetable oil 10g
Sesame oil 5g
Light soy sauce 3g
White pepper powder 2g
Minced spring onion 5g
Minced ginger 5g

Method
1. Season the pork mince with all seasonings except for the broth. Stir the pork mince in one direction until the meat is sticky.
2. In a wok, heat the supreme broth up to about 30°C, grasp some meat with one hand, make a fist, squeeze a meatball out from the top of the fist, scoop it into the broth one by one. The meatball should be 2.5-3cm diameter in size.
3. Remove scum, simmer the meatballs on low heat until cooked. Colander them out to leave them in a soup bowl.
4. Bring the soup to a boil in the pot, add black ear fungus and season with salt.
5. Pour the soup to the soup bowl, garnish with finely chopped coriander.
6. Serve.

Feature of the Dish
This dish is featured with palatable tender meatballs and umami soup.

Chinese Culinary Techniques

SHUAN LAMB (涮羊肉)

Ingredients
Finely sliced Lamb 2,000g
Rehydrated mung beans noodles 250g
Frozen tofu 150g
Fresh or rehydrated dried Mushrooms 200g
Greens and vegetables

Ingredients for base soup
Water or chicken stock 2L
Rehydrated dried shrimps 30g
Ginger 30g
Spring onion 30g

Seasoning for dipping sauce
Finely chopped spring onions 100g
Finely chopped coriander 50g
Rice vinegar 75g
Dried-chili oil 100g
Fermented Chinese chive flower sauce 50g
Shrimp sauce 100g
Sesame paste 100g
Red fermented bean curd 50g
Salt 10g
Cooking wine 50g
Sesame seeds 100g
Vegetable oil as appropriate

Method

1. To make the base soup.
   In northern regions of China, clear chicken stock or water seasoned with ginger, spring onions are the common choice of the base soup. Dried shellfish such as shrimps are often used to add an umami taste to the soup.

2. To make the northern style dipping sauce.
   In a small mixing bowl, add sesame oil into sesame paste to make it runny, mash red fermented bean curd cubes with its brine. Mix them well in a separate mixing bowl. Add fermented Chinese chive flour sauce, shrimp sauce, cooking wine, mix until the sauce is smooth. The mixture is served as the base of northern style dipping sauce. Finely chopped spring onions, coriander, chili oil, vinegar, salt, sesame seeds can be added according to individual taste.

3. To prepare ingredients.
   Mung mean noodles needs to be rehydrated so do dried mushrooms, frozen tofu should be defrosted at room temperature.

4. To enjoy, scald the lamb in the base soup, take it out to dip in the dipping sauce once it turns grey which is a sign of being cooked.

Feature of the Dish
A balanced meal with a wide range of ingredients to choose from. Enjoy it with your family and friends in a fun way.

Part Three

SALT BAKED CHICKEN WINGS

Ingredients
Chicken wings 12

Seasonings
Sand ginger power 10g
Chopped spring onion 10g
Coriander 25g

Rock salt 2,500g
Table salt 12g
Star anise powder 2g
Sesame oil 2g
Sliced ginger 3g
Sliced spring onion 3g
Vegetable oil 50g

Method
1. Marinate chicken wings with table salt 2g, spring onions, ginger, and star anise powder.
2. Mix 30g of vegetable oil, sesame oil, and table salt 6g in a small bowl. Brush the mixture onto a piece of tin foil that is big enough to wrap 12 chicken wings. Lay another piece of foil on top, place seasoned chicken wings in and wrap them up first with the non-greased tin foil and then with the greased foil. Set it aside.
3. Pre-heat a clean wok on high heat, stir fry the rock salt, when the salt is heated up, take ¼ of salt out. Lay the wrapped chicken wings in the wok on top of the salt, cover it with the ¼ of the salt. Place the lid on, Ju on low heat for about 20 minutes until cooked.
4. Remove the chicken wings from the warp.
5. Garnish with coriander and serve with the dipping sauce.

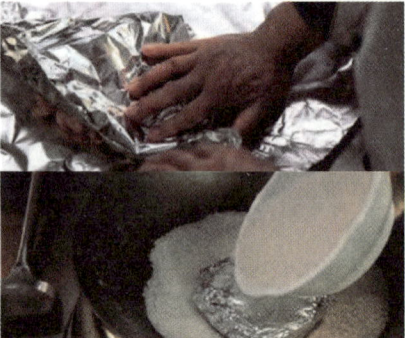

To make the dipping sauce
1. Pre-heat the wok on low heat, stir fry 4g of table salt.
2. When the wok gets hot, add in sand ginger power, stir well, stop cooking when they are aromatic, place in a small bowl.
3. Add 15g of vegetable oil into the small bowl. Evenly divide the sauce into 3 portions.

Feature of the Dish
Salt baked chicken wings are moist, smooth, aromatic, and savoury. The dipping sauce adds a distinctive flavour.

WOK ROAST CHICKEN WINGLETS

Ingredients
Winglets 18

Seasonings
Finely chopped ginger 5g
Finely chopped spring onion 5g
Salt 3g
Cooking wine 15g
Light soy sauce 1g
Sugar 25g
Lea & Perrins sauce 30g
Tomato sauce 15g
Vegetable oil 1,000g
White pepper powder 4g

Method
1. Score twice on each side of a winglet to let marinade sets in easier.
2. Blanch the winglets to reduce the gamey taste, scoop out from the boiling water in a colander, let them drain well.
3. Add 1g of light soy sauce in the winglets to enhance the colour.
4. Pour 1,000g of vegetable oil into a clean wok, bring it to 210°C, carefully lower the winglets in, remove them when the thin soy sauce coating becomes firm and drain them in the colander.
5. Place the empty wok back to stove, stir fry the spring onions and ginger until aromatic, stir in tomato sauce, cook briefly, add in the winglets, season the dish with cooking wine, light soy sauce, sugar, salt, Lea & Perrins sauce, and white pepper powder. Add enough water to submerge the winglets.
6. Bring it to the boil, remove scum, simmer the winglets on low heat until the broth thickened, and the winglets are soft and tender. Now, increase the heat level to medium to further thicken the broth, add a splash of oil to give the dish a glossy appearance.
7. Dish out and serve.

Feature of the Dish
This dish has a glossy dark brown colour, a sweet and salty taste with a hint of piquancy.

Part Three

LU DUCK GIZZARDS

Ingredients

Duck gizzards 1,000g
Celery 150g
Coriander 45g
Carrots 200g
Onions 30g
Ginger 30g
Green peppers 100g
Dried scallop 80g
Dried shrimps 40g
Dry cured ham 40g
Water 4,000g

Seasonings

Light soy sauce 90g
Dark soy sauce 35g
Maggi seasoning soy sauce 15g
Fish sauce 30g
Lea & Perrins sauce 15g
Soy sauce for fish 15g
Rock sugar 40g
Cooking wine 60g
Black peppercorns 10g
Salt 10g
Sha ren (cructus amomi) 8g
White cardamom 10g
White angelica root 15g
Cao guo (Tsao-ko cardamom) 5g
Fennel seed 8g
Luo han guo 15g
Sichuan peppercorns 6g
Star anise 8g
Bay leaf 3g
Lemon grass 5g

Method

1. Wash the duck gizzards in cold water and leave them to soak for 20 minutes.
2. Boil 4,000g of water in the wok, plunge the duck gizzards in the boiling water, blanch them to reduce the gamey taste, take them out and leave them in cold water to cool down. This step will also help to firm the texture.
3. To make the red broth,
    a. Add the black peppercorns 10g, sha ren (fructus amomi) 8g, white cardamom 10g, white angelica root 15g, cao guo (tsao-ko cardamom) 5g, fennel seed 8g, Luo han guo 15g, Sichuan peppercorns 6g, star anise 8g, bay leaf 3g, and lemongrass 5g in a muslin food bag. Tie the bag up and leave it in the pot with 4,000g water.

    b. Add dried scallop 80g, dried shrimps 40g, and dry-cured ham 40g into the second muslin bag.

      Tie it up and plunge into the pot.

   c. Drop in Celery 150g, coriander 45g, carrots 200g, onions 30g, ginger 30g, and green peppers 100g to the pot.

   d. Season with light soy sauce 90g, dark soy sauce 35g, maggi seasoning soy sauce 15g, fish sauce 30g, Lea & Perrins sauce 15g, soy sauce for fish 15g, rock sugar 40g, and cooking wine 60g.

   e. Bring water to the boil and cook on high heat for 20 minutes.

4. Cook duck gizzards in the red broth on low heat for 40 minutes.
5. Remove the duck gizzards from the red broth.
6. Brush a lay of oil onto the duck gizzards.
7. Slice the duck gizzards, garnish with cooked sweet peas and carnation petals and serve.

Feature of the Dish

This dish has a glossy red brown colour, a tender and moist texture with a strong aroma.

Part Three

LU CHICKEN LIVERS (卤鸡肝)

Ingredients
Chicken Livers 750g
Coriander 30g
Celery 20g
Onion 20g
Ginger 20g
Water 2,000g

Seasonings
Cooking wine 60g
Salt 20g
Sichuan peppercorns 6g
Cao guo (Tsao-ko cardamon) 5g
White angelica root 15g
Fennel seeds 5g
Bay leaf 2g
White cardamon 8g

Method
1. Leave the chicken livers in cold water to soak for 20 minutes to remove the blood.
2. Boil water (enough to cover the livers) in a wok, blanch the livers to reduce the gamey taste. Filter the livers out and rainse them in cold water.
3. Add Sichuan peppercorns 6g, cao guo cardamom (Tsao-ko cardamon) 5g, white angelica Root 15g, fennel seeds 5g, bay leaf 2g, and white cardamon 8g in a muslin food bag.
4. Tie the bag up and leave it in a pot with 2,000g of water. Add cooking wine 60g, salt 20g, coriander, celery, onion, and ginger. Bring water to the boil, simmer for 10 minutes on high heat.
5. Plunge the livers in, simmer on low heat for 15 minutes.
6. Remove the pot from heat, infuse the livers in the broth for another 15 minutes.
7. Scoop the livers out, place on a plate, and brush a layer of sesame oil.
8. Slice the chicken liver.
9. Plating and serve.

Feature of the Dish
The chicken livers are tender, savoury, umami, and aromatic.

Section 3 Vegetable Recipes

FIVE-FLAVOURED CUCUMBERS

Ingredients
Cucumbers 400g

Seasonings
Salt 8g

Sugar 30g
Dried chili 3g
Rice vinegar 30g
Ginger juliennes 10g
Rehydrated shiitake mushrooms 5g

Method
1. Wash the cucumbers, deseed, and cut into strips in uniform size.
2. Shred the shiitake mushrooms and break the dried chili into smaller pieces.
3. Mix the cucumber strips with 8g salt in a mixing bowl and marinate for 20 minutes.
4. Gently squeeze the cucumber strips to remove the emitted liquid. Place them in a clean mixing bowl.
5. Add the mushrooms, ginger, dried chili, rice vinegar, and sugar into the mixing bowl. Mix them well with the cucumber.
6. Seal the mixing bowl with cling film, store it in the fridge for 30 minutes.
7. Plating and serve.

Feature of the Dish
This dish has a dark green colour. It is crunchy with a well-balanced compound taste of sour, sweet, spicy, and salty and a ginger aroma.

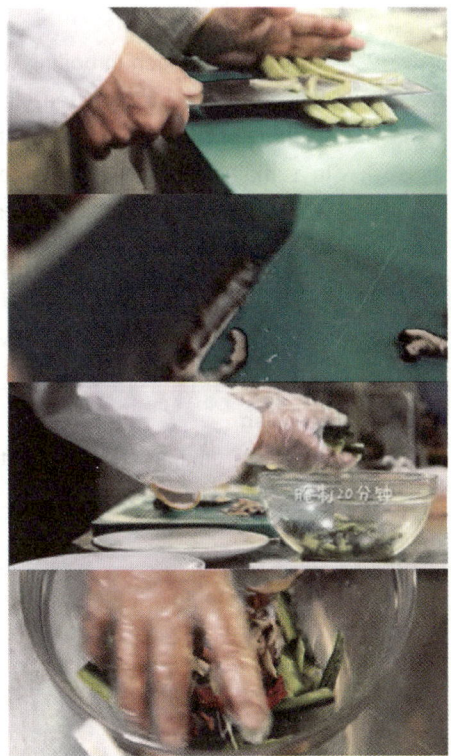

SPINACH WITH GINGER DRESSING

Ingredients
Spinach 500g
Baby ginger 50g
Coriander 20g

Seasonings
Sesame oil 15g
Salt 7g
Fragrant vinegar 20g
Soy sauce for fish 20g

Method
1. Wash the spinach and coriander, cut the spinach into 3cm long sections and the coriander into 1cm long segments.
2. Blanch the spinach in boiling water, cool it in cold water, and drain it by squeezing gently.
3. Peel the baby ginger and chop it finely.
4. In a small mixing bowl, combine the minced ginger with fragrant vinegar, soy sauce for fish, salt, and sesame oil to complete the dressing.
5. Pour the ginger dressing onto the spinach, combine them well.
6. Serve immediately.

Feature of the Dish
The dish has a rich aroma, a refreshing taste and a vibrant green color.

SWEET AND SOUR SHANGHANG WHITE RADISH

Ingredients
Chinese white radish (mooli) 600g

Seasonings
Salt 10g

Sugar 45g
Light soy sauce 30g
Apple cider vinegar 20g
Sesame oil 5g

Method
1. Wash and peel the radish. Cut it in quarters lengthwise.
2. Slice the radish into half-splits, each piece should be about the thickness of a coin, place them in a mixing bowl.
3. Add salt and sugar, seal the bowl with cling film, marinate in the fridge for 1 hour.
4. Gently squeeze the white radish to remove emitted liquid and leave them in a mixing bowl, season with light soy sauce, apple cider, and sesame oil.
5. Seal the blow again with cling film and leave it to marinate in the fridge for 20 minutes.
6. Plating and serve.

Feature of the Dish
The radish has an appetising golden yellow colour, a crunchy and chewy texture with a combined taste of sweet, sour, and salty.

Part Three

CABAGE PICKLES

Ingredients
Cabbage 500g
Carrots 120g
Green sweet peppers 100g
Onion 100g

Seasonings
Curry powder 40g
Bay leave 2g
Dried chili peppers 8g
Salt 6g
Sugar 30g
Rice vinegar 30g

Method

1. Wash the cabbage, halve it lengthwise, remove the core, cut it into the elephant-eye (Rhomboid) shaped pieces.
2. Cut the green peppers, carrots, and onion into the same shape.
3. Mix the cabbage with curry power in a mixing bowl, leave it to marinate. Put the rest of the vegetables in the bowl.
4. Bring water (enough to cover all the ingredients) to the boil in wok, season with bay leaves and the remaining curry power. Boil for 2 minutes.
5. Blanch the vegetables in the boiling water, filter them out to drain.
6. Mix the blanched vegetables with dried chilli peppers, salt, sugar, and rice vinegar in a mixing bowl.
7. Cover the bowl with cling film and store it in the fridge for 24 hours.
8. Plating and serve.

Feature of the Dish

The dish has a bright assorted colour and a well-balanced sweet and sour taste. It is refreshing and crunchy with a distinctive curry aroma.

STIR-FRIED LILY BULB AND CELERY

Ingredients
Celery stalks 250g
Eatable lily bulbs (fresh or rehydrated from the dried) 20g

Seasonings
Vegetable oil 15g
Salt 5g
Corn starch 1g

Method
1. Peel the celery stalks to remove the tough skin, slice them obliquely into 2.5mm fine slices.
2. Add some water into the wok, bring it to the boil, add in salt 3g and oil 5g, blanch the lily bulbs briefly, rinse them with cold water. Blanch the celery stalk, remove them from the wok when the colour turns bright green, let them drain.
3. Put the empty wok back onto stove, add 6g of oil, stir in celery and lily bulb, season with salt, add in corn starch solution and then a splash of oil 4g.
4. Dish out and serve.

Feature of the Dish
The dish has a delightful colour combination of white and green. It is refreshing, savoury and umami.

Part Three

WOK TA TOFU

Ingredients
Fresh tofu 250g

Seasonings
Corn starch solution 30g
Cooking wine 10g
Salt 6g

Sesame oil 10g
Finely chopped onion 10g
Vegetable oil 100g
Minced ginger 5g
Broth 80g
Eggs 2
Flour 50g

Method
1. Slice the tofu into 5cm long, 4cm wide, 0.8cm thick pieces.
2. Spread the tofu pieces on a plate, season with salt 4g, chopped spring onions 4g, minced ginger 2g, and cooking wine 4g. Leave the tofu to marinate.
3. Preheat wok on medium heat, pour 100g of vegetable oil in, bring it to 70°C. While waiting, crack and beaten 2 eggs in a mixing bowl. In a separate mixing bowl, place 50g of dry flour.
4. When the oil is ready, pad flour on both sides of the tofu, and then dip it in the beaten egg, carefully lay it in the oil.
5. Repeat this process until all the pieces of tofu are coated and laid in the oil. Toss the wok in circles to prevent the tofu from sticking to the wok.
6. Carefully flip the tofu to the other side, fry until the bottom side turns golden.
7. Introduce clear broth, spring onions 6g, ginger 3g, cooking wine 6g, salt 2g, and liquid corn starch to the wok, bring the broth to the boil.
8. Sprinkle a splash of sesame oil when the broth reduced.
9. Dish out and serve.

Feature of the Dish
The tofu has a golden colour. It is tender with a rich and savoury taste.

MAPO TOFU

Ingredients
Fresh tofu 200g
Minced beef 75g
garlic greens 15g (garnish)

Seasonings
Minced garlic 5g
Minced spring onion 5g
Minced ginger 5g
Salt 2g

Pixian broad bean paste 20g
Fermented black beans 15g
Red chili oil 10g
Light soy sauce 10g
Liquid corn starch 5g
Clear broth
Chili powder 3g
Sichuan peppercorn powder 3g
Vegetable oil 75g

Method
1. Cut tofu into 2.5 x 2.5 x 2.5cm dices, slice the garlic green obliquely.
2. Bring water to a boil in the wok, add a teaspoon of salt, plunge the tofu dices in for blanching. Pour the hot water together with the tofu into a deep bowl, leave the tofu to soak in the hot water.
3. Add vegetable oil to a pre-heated wok, quickly stir fry the minced beef, remove it from the wok when it turns white. Leave the beef mince in a small bowl, set aside.
4. Stir fry minced garlic, ginger, and spring onions with a small amount of oil until aromatic, rapidly stir in broad bean paste, fermented black bean, and chili powder. Add clear broth to the mixture. Bring it to the boil.
5. Plunge the drained tofu dices into the wok, season with soy sauce and salt. Keep tossing the wok.
6. Cook until broth reduced, add in cornflour liquid to thicken the broth while continue with the wok tossing to avoid food getting stuck on the wok.
7. Add in garlic greens and Sichuan pepper powder. Toss the wok to mix well.
8. Dish out and serve.

Feature of the Dish
This tofu dish is savoury, soft, tender, and smooth. Coated with dark-red-coloured broth, this dish has a strong fermented paste aroma. It is spicy, umami with a mouth-numbing sensation.

Part Three

Section 4 Noodles, Dumplings, and Flatbread Recipes

HOMEMADE FLATBREAD

Ingredients
Strong flour 1,000g
Vegetable oil 50g
Cold water 75g

Seasonings
Salt 20g

Method

1. Forming the dough
   Place the flour on the worktable surface and make a hole in the middle with your hand, add some of the water into the hole. Slowly incorporate the surrounding flour with the water reaching a stage that is called "ears of the wheat" which is when the flour comes together with the water to form a breadcrumb stage. This is when you can knead until a dough is formed. Be careful not to add too much water as you will flood the dough. Continue to form the dough, adding more water if necessary, constantly feel the dough making sure that is not to dry or firm. Once the dough has formed, continue to develop the gluten in the dough by kneading until the surface is smooth. Leave it to rest for 15 minutes with a damp cloth over the top of the dough to prevent it from dry and forming a crust on the surface.

2. Divide and forming the flatbread
   Once the dough has rested for 15 minutes, divide it by cutting the dough from left to right to form long strips of dough.
   Roll each piece of dough into strips around 10cm, brush on vegetable oil and sprinkle with a little bit salt. From one end, began to roll the dough up into a cylinder, then rest for 2 minutes before rolling it out again into a circle that is 1cm thick.

3. Cooking
   Preheat the flat bread maker to 180°C, brush one side of the flatbread with oil, then place it on flatbread maker, begin cooking. When the flatbread turns golden brown, brush the uncooked surface with vegetable oil and turn over to finish cooking. Ensure that both sides of the flatbread are golden brown and cooked evenly.

Chinese Culinary Techniques

FLATBREAD WITH MEAT FILLING

Ingredients
Strong flour 500g
Pork mince 250g
Cold water 250g

Seasonings
Sesame oil 30g
Cooking Wine 20g
Salt 20g
Soy sauce 20g
Oyster sauce 10g
Minced spring onions 10g
Minced ginger 10g

Method

1. Forming the dough
   Place the flour on the worktable surface and make a hole in the middle with your hand, add some of the water into the hole. Slowly incorporate the surrounding flour with the water reaching a stage that is called "ears of the wheat" which is when the flour comes together with the water to form a breadcrumb stage. This is when you can knead until a dough is formed. Be careful not to add too much water as you will flood the dough. Continue to form the dough, adding more water if necessary, constantly feel the dough making sure that it is not too dry or firm. Once the dough has formed, continue to develop the gluten in the dough by kneading until the surface is smooth. Leave it to rest for 15 minutes with a damp cloth over the top of the dough to prevent it from dry and forming a crust on the surface.

2. Preparing the pork mince filling
   Place the pork mince in a bowl and add the salt, soy sauce, oyster sauce, cooking wine, onion, ginger and stir them evenly, add a small amount of the cold water and stir. Add more water to reach the correct consistency and finally add sesame oil and mix well.

3. Forming the flatbread
   Once the dough has rested for 15 minutes, flatten it out with a rolling pin to form a rectangular sheet, place the pork mince in the center of the dough leaving an inch of dough around all the sides. Roll the dough from the top to the bottom keeping it in a rectangular shape, then place on a wooden tray. On the work surface, use the rolling pin to press the flatbread into a flat uniformed shape.

4. Cooking
   Preheat the flat bread maker to 180°C, brush vegetable oil onto one side of raw meatloaf and place onto the flat bread maker. Cook until golden brown then brush with vegetable oil on the uncooked side of the meatloaf and turn to cook again until golden brown. Once cooked, remove from the flat bread maker and cut into strips 6cm wide to serve.

Part Three

SPRING ROLLS

Ingredients
Pork 250g
Spring roll sheets 500g
Chinese chive (garlic chive) 500g
Spring onion 9g
Winter bamboo shoots 150g
Sea cucumber (operational) 30g
Soy sauce 15g
Minced ginger 15g
Minced spring onions 15g
Egg 1
Peanut oil for frying 500ml

Seasonings
Cooking Wine 10g
Flour 30g
Salt 8g
Corn starch solution 20g

Method

1. Preparing the filling

   Begin to slice the pork to create whole thin cuts, then slice into long thin rectangular sections. Take your winter bamboo shoots and repeat the same process to create thinner matchsticks. Then roughly chop the Chinese chive and slice the sea cucumber. Then finely chop spring onions and ginger.

   To complete the spring roll filling, place the pork in a bowl and blend together with half an egg yolk, then coat in flour. Begin to shallow fry off the pork in a wok and then set to one side to drain. Then stir fry minced ginger and spring onions until they are aromatic. Return the pork to the wok. Add the bamboo shot shreds, sea cucumber, cooking wine, salt, soy sauce, and corn starch solution to thicken the mixture. Once it has thickened and reduced, remove from the heat.

   Prepare the water and flour into a paste to use to stick the spring roll pastry together later.

2. Preparing the spring rolls

   Place a suitable amount of the filling into the centre of the spring roll sheet, ensure that the square has one of the edges pointing in front of you. Begin to roll from the bottom to the top, leaving about 5cm of the sheet unrolled. Cover the left and right sides of the roll with the premade flour and water solution and fold each end into the centre. Finally close the roll with the water and flour solution again on the 5cm part of unrolled pastry and complete the roll.

3. Cooking

   Preheat the peanut oil to 210°C - 240°C in a wok and begin to fry the spring rolls, once they are golden brown remove and drain prior to serving.

Chinese Culinary Techniques

SANXIAN GRAVY NOODLES

Ingredients

Strong flour 500g
Cold water 250g
Shredded cucumber 50g
Shredded carrot 50g
Minced pork 100g
Sea cucumber (optional) 50g
Sweat peas 50g
Soybean 50g
Eggs 3
Prawns 50g
Bamboo shoot 50g
Rehydrated dried Shiitake Mushroom 20g
Rehydrated eatable dried day lily 20g
Rehydrated dried tofu sticks 20g
Black ear mushroom 20g
Gluten 20g
Five spice dried tofu 20g
Bean sprouts 20g
Spinach 50g

Seasonings

Salt 5g
Soy sauce 20g
Cooking wine 20g
Star anise 2

Method

1. Forming the dough
   Place the flour on the worktable surface and make a hole in the middle with your hand, add some of the water into the hole. Slowly incorporate the surrounding flour with the water reaching a stage that is called "ears of the wheat" which is when the flour comes together with the water to form a breadcrumb stage. This is when you can knead until a dough is formed. Be careful not to add too much water as you will flood the dough. Continue to form the dough, adding more water if necessary, constantly feel the dough making sure that is not too dry or firm. Once the dough has formed, continue to develop the gluten in the dough by kneading until the surface is smooth. Leave it to rest for 15 minutes with a damp cloth over the top of the dough to prevent it from dry and forming a crust on the surface.

2. Making the noodles
   Once the dough has rested for 15 minutes, roll it out into a large rectangle and flour to prevent it from sticking together. To achieve this process, you can roll the dough out, then use a rolling pin to roll the dough up. Pull the rolling pin out of the dough and widen the dough by pressing down on the dough repeatedly with the rolling pin. Repeat this vertically and horizontally, then re-roll the dough out to complete the required seize. Concertina the dough back and forth so the dough is folded up upon itself and is no wider than 15cm. Now you can begin to cut the dough into 6mm wide noodles, again sprinkle with more flour to prevent them from sticking together prior to cooking.

3. Preparing the vegetable, meat and prawn

   Begin to prepare by slicing the gluten, five spice dried tofu, rehydrated dried tofu sticks, black ear mushroom, eatable yellow day lily, shiitake mushroom, the bamboo shoot. Leave each one prepared on a plate to begin cooking.

   Then slice the pork into thin strips along with the sea cucumber and split the prawns in half longways and cut the spinach into sections.

   Separately blanch all the soybeans, green peas, bean sprouts, spinach and carrots in boiling water and leave to one side on a plate, then fry off the eggs and leave for later for the final dish. Coat the prawns in a corn starch solution and egg yolk and leave to aside.

4. Cooking the gravy

   Add vegetable oil to your wok and two pieces of star anise, heat the oil and then add the strips of pork and begin to cook. Add the cooking wine and soy sauce then add water and begin to cook. Remove the star anise and bring water to the boil, add the pre prepared sliced gluten, five spice dried tofu, rehydrated dried tofu sticks, black ear mushrooms, eatable yellow day lily, shiitake mushrooms, the bamboo shoots and allow to cook, ensure you season correctly with salt. Add a little corn starch solution to thicken and carry on with the cooking until the corn starch has completely cooked out and place the gravy in a bowl ready to serve.

   Cook the prepared prawns in vegetable oil and leave to one side. Boil water and add your noodles, once cooked remove and sieve off the excess water then place in a serving bowl. Now begin to build your dish by adding to the surface of the noodles the spinach, bean sprouts, cucumber, soybeans, green peas and precooked scrambled egg and carrot. To finish ladle on the pork mixture on top vegetable in the centre of the bowl and place the precooked prawns on top to serve.

Chinese Culinary Techniques

NOODLES WITH SOYBEAN PASTE (ZHA JIANG MIAN)

Ingredients
Strong flour 500g
Cold water 250g
Bean sprouts 20g
Sweet peas 50g
Soybean 50g
Spinach 50g
Shredded carrot 50g
Shredded cucumber 50g
Eggs 3

Seasonings
Salt 5g
Soy sauce 20g
Cooking wine 20g
Sweet flour sauce (Tian Mian Jian) 20g
Minced pork 100g
Shiitake mushrooms 10g
Fermented soybean 10g
Star anise 2
Corn starch solution (for thickening) 10g

Method

1. Forming the dough
   Place the flour on the worktable surface and make a hole in the middle with your hand, add some of the water into the hole. Slowly incorporate the surrounding flour with the water reaching a stage that is called "ears of the wheat" which is when the flour comes together with the water to form a breadcrumb stage. This is when you can knead until a dough is formed. Be careful not to add too much water as you will flood the dough. Continue to form the dough, adding more water if necessary, constantly feel the dough making sure that it not too dry or firm. Once the dough has formed, continue to develop the gluten in the dough by kneading until the surface is smooth. Leave it to rest for 15 minutes with a damp cloth over the top of the dough to prevent it from dry and forming a crust on the surface.

2. Making the noodles
   Once the dough has rested for 15 minutes, roll it out into a large rectangle and flour to prevent it from sticking together. To achieve this process, you can roll the dough out, then use a rolling pin to roll the dough up. Pull the rolling pin out of the dough and widen the dough by pressing down on the dough repeatedly with the rolling pin. Repeat this vertically and horizontally, then re-roll the dough out to complete the required seize. Concertina the dough back and forth so the dough is folded up upon itself and is no wider than 15cm. Now you can begin to cut the dough into 6mm wide noodles, again sprinkle with more flour to prevent them from sticking together prior to cooking.

3. Vegetable preparation
   Cut the shiitake mushrooms into small dices, roughly chop the spinach, shred the cucumber, and leave them to one side. Then blanch separately the sweet peas, soybeans, beansprouts, carrots, and spinach. Scramble the eggs in vegetable oil and leave the cucumber raw. Leave all the items on a

plate for use later whilst cooking the sauce.

4. Cooking the noodle sauce

In your wok add vegetable oil and two pieces of star anise, heat and then remove the star anise. Add shiitake mushrooms dices, fermented soybean, the pork mince, cooking wine, and cook off. When it is aromatic, add sweet flour sauce and soya sauce, continue to cook. Bring to a rapid boil and cook until there is evidence of oil and then empty into a bowl. To serve boil the noodles and drain them off and place them in a bowl, arrange the precooked off the sweat peas, soybeans, beansprouts, carrots and spinach, scrambled eggs, and the cucumber on the surface of the noodles and ladle the fried sauce to the centre of the dish.

Chinese Culinary Techniques

NOODLES IN PICKLED MUSTARD GREEN GRAVY

Ingredients
Strong flour 500g
Cold water 250g
A pinch of salt
Pork 500g
Pickled mustard green stems 100g
Celtuce (also known as Chinese lettuce) 50g

Seasonings
Salt 5g
Soy sauce 20g
Cooking wine 20g
Sichuan peppercorn 5g
Star anise 5g
Spring onions 10g
Ginger 10g

Method
1. Forming the dough
   Place the flour on the worktable surface and make a hole in the middle with your hand, add some of the water into the hole. Slowly incorporate the surrounding flour with the water reaching a stage that is called "ears of the wheat" which is when the flour comes together with the water to form a breadcrumb stage. This is when you can knead until a dough is formed. Be careful not to add too much water as you will flood the dough. Continue to form the dough, adding more water if necessary, constantly feel the dough making sure that is not too dry or firm. Once the dough has formed, continue to develop the gluten in the dough by kneading until the surface is smooth. Leave it to rest for 15 minutes with a damp cloth over the top of the dough to prevent it from dry and forming a crust on the surface.
2. Making the noodles
   Once the dough has rested for 15 minutes roll it out into a large rectangle and flour to prevent it from sticking together. To achieve this process, you can roll the dough out, then use a rolling pin to roll the dough up. Pull the rolling pin out of the dough and widen the dough by pressing down on the dough repeatedly with the rolling pin. Repeat this vertically and horizontally, then re-roll the dough out to complete the required seize. Concertina the dough back and forth so the dough is folded up upon itself and is no wider than 15cm. Now you can begin to cut the dough into 6mm wide noodles, again sprinkle with more flour to prevent them from sticking together prior to cooking.
3. Process the ingredients
   Shred 5g spring onions, pickled mustard greens, Chinese lettuce, and pork. Chop the rest of the spring onions and ginger.
4. Cook the noodles
   Bring water into the boil in a wok and add the noodles, remove when cooked and place in cold water

to cool and then drain the water off. Set aside for later use.

5. Cook the Noodle Soup

Stir fry the chopped spring onions and ginger in wok until aromatic, add in the pork shreds, cooking wine and soy sauce. Continue to stir. Add water, Sichuan peppercorns and star anise. Bring it to the boil, plunge in the pickled mustard greens and simmer for 10 minutes. Place the cooked noodles in a noodle colander, rinse with the hot soup over the wok. Drain and place in a bowl, top up the noodles with the soup then garnish with the shredded spring onions and serve.

Chinese Culinary Techniques

NOODLES IN TOMATO EGG SAUCE

Ingredients
Strong flour 500g
Cold water 250g

Seasonings
Tomato 200g
Eggs 3
Salt 5g
Finely chopped spring onions 10g
Finely chopped ginger 10g
Coriander for garnishing

Method

1. Forming the dough
   Place the flour on the worktable surface and make a hole in the middle with your hand, add some of the water into the hole. Slowly incorporate the surrounding flour with the water reaching a stage that is called "ears of the wheat" which is when the flour comes together with the water to form a breadcrumb stage. This is when you can knead until a dough is formed. Be careful not to add too much water as you will flood the dough. Continue to form the dough, adding more water if necessary, constantly feel the dough making sure that is not too dry or firm. Once the dough has formed, continue to develop the gluten in the dough by kneading until the surface is smooth. Leave it to rest for 15 minutes with a damp cloth over the top of the dough to prevent it from dry and forming a crust on the surface.

2. Making the noodles
   Once the dough has rested for 15 minutes roll it out into a large rectangle and flour to prevent it from sticking together. To achieve this process, you can roll the dough out, then use a rolling pin to roll the dough up. Pull the rolling pin out of the dough and widen the dough by pressing down on the dough repeatedly with the rolling pin. Repeat this vertically and horizontally, then re-roll the dough out to complete the required seize. Concertina the dough back and forth so the dough is folded up upon itself and is no wider than 15cm. Now you can begin to cut the dough into 6mm wide noodles, again sprinkle with more flour to prevent them from sticking together prior to cooking.

3. Cooking the noodles
   Prepare two tomatoes by blanching them and then remove the skins. Slice the tomatoes into wedges. Cook the noodles in boiling water in a wok, remove once cooked and cool in cold water then drain in a colander to remove excess water. Set aside for later use.
   In a wok, add a small amount of vegetable oil and begin to fry off the chopped spring onions and ginger until aromatic, add tomato wedges and salt in and continue to stir. As the tomatoes begin to break down, add three ladles of water. Continue to cook on a rapid boil and add the beaten eggs. When eggs become solid, dip the noodles into the tomato sauce to heat them trough and place them in a serving bowl. To serve, ladle more of the tomato broth on the noodles and garnish in the centre with chopped coriander.

Part Three

YIFU NOODLES

Ingredients
Strong flour 500g
Eggs 3
Pork tenderloin 100g
Shelled prawns 80g
Sea cucumber 50g
Rehydrated dried bamboo shoots 80g
Shiitake mushroom 50g
Pak choi 50g

Seasonings
Ginger 10g
Spring onions 10g
Soy sauce 10g
Cooking wine 10g
Salt 5g

Method

1. Forming the dough
   Place the flour on the worktable surface and make a hole in the middle with your hand, add three eggs into the hole. Slowly incorporate the surrounding flour with the eggs reaching a stage that is called "ears of the wheat" which is when the flour comes together with the eggs to form a breadcrumb stage. now you can knead until a dough is formed. Be careful not to add too much egg as you will flood the dough. Continue to form the dough, adding more egg if necessary, constantly feel the dough making sure that it is not too dry or firm. Once the dough has formed, continue to develop the gluten in the dough by kneading until the surface is smooth. Leave it to rest with a damp cloth over the top of the dough to prevent it from drying and forming a crust on the surface.

2. Making the noodles
   Once the dough has rested for 15 minutes, roll it out into a large rectangle and flour to prevent it from sticking together. To achieve this process, you can roll the dough out, then use a rolling pin to roll the dough up. Pull the rolling pin out of the dough and widen the dough by pressing down on the dough repeatedly with the rolling pin. Repeat this vertically and horizontally, then re-roll the dough out to complete the required seize. Concertina the dough back and forth so the dough is folded up upon itself and is no wider than 15cm. Now you can begin to cut the dough into 6mm wide noodles again sprinkle with more flour to prevent them from sticking together prior to cooking.

3. The 1st stage of cooking the noodles
   Blanch the noodles in boiling water in your wok, remove and cool in cold water then drain in a colander to remove excess water. Ensure you remove the water from the noodles as the next stage involves cooking in oil. Heat your oil in a saucepan and carefully place the noodles into the oil. After a few minutes remove the noodles once they are golden and place on a plate for later use.

4. Preparing the sauce

    Slice the pork into thin slices and then into strips, split each prawn in half, slice each mushroom, then slice the rehydrated dried bamboo shoots into whole thin slices and then cut into strips and cut the sea cucumber into thin strips as well.

    Place the pork meat into a bowl and add water, half an egg yolk and cover with a few pinches of flour and bind all the ingredients together. Leave to one side and repeat the process again with the prawns.

5. The 2nd stage of cooking the noodles

    In this stage the noodles will be cooked again in boiling water. Using a wok to boil water and make sure it is seasoned with salt and add the fried noodles. Remove them as soon as they are free and move around as the frying process sets them solid.

    In a wok, begin to fry off the pork and then the prawns, set them to one side. Empty the wok and add water, bring to the boil then blanch the mushrooms, rehydrated dried bamboo shoots, sea cucumber and pak choi. Now begin to stir fry minced ginger and spring onions until aromatic, add the precooked shredded pork, soy sauce, salt and then top up with two ladles of water. Now add the pre blanched mushrooms, rehydrated dried bamboo shoots, sea cucumber and pak choi. Season and add the noodles, pak choi, prawns and serve.

JIAO STIR-FRIED NOODLES

Ingredients
Strong flour 500g
Cold water 250g
Eggs 3
Pork 100g
Spinach 100g
Vegetable oil 50g

Seasonings
Cooking wine 20g
Soy sauce 20g
Salt 5g
Spring onions 10g
Ginger 10g

Method

1. Forming the dough
   Place the flour on the worktable surface and make a hole in the middle with your hand, add some of the water into the hole. Slowly incorporate the surrounding flour with the water reaching a stage that is called "ears of the wheat" which is when the flour comes together with the water to form a breadcrumb stage. This is when you can knead until a dough is formed. Be careful not to add too much water as you will flood the dough. Continue to form the dough, adding more water if necessary, constantly feel the dough making sure that is not to dry or firm. Once the dough has formed, continue to develop the gluten in the dough by kneading until the surface is smooth. Leave it to rest with a damp cloth over the top of the dough to prevent it from drying and forming a crust on the surface.

2. Making the noodles
   Once the dough has rested for 15 minutes, roll it out into a large rectangle and flour to prevent it from sticking together. To achieve this process, you can roll the dough out then use a rolling pin to roll the dough up. Pull the rolling pin out of the dough and widen the dough by pressing down on the dough repeatedly with the rolling pin. Repeat this vertically and horizontally, then re-roll the dough out to complete the required seize. Concertina the dough back and forth so the dough is folded up upon itself and is no wider than 15cm. Now you can begin to cut the dough into 6mm wide noodles, again sprinkle with more flour to prevent them from sticking together prior to cooking.

3. Steaming the noodles (The 1st stage of cooking the noodles)
   Put the noodles into the steamer baskets and sit them in a wok with boiling water. Ensure that the noodles are spaced evenly in the baskets, so the noodles cook evenly. Steam for about 20 minutes and then place the noodles on the work surface to dissipate, flour them again and place aside for later.

4. Preparing the sauce
   To prepare the sauce firstly shred the pork by fanning the pork out into sheets and then cut them

into strips. Take the spinach and cut roughly and chop both the spring onions and ginger into small pieces.

5. The 2nd stage of cooking the noodles

   Continue cook the base of the noodles by scrabbling three of the eggs. Remove and set aside for later use. Now, clean the wok, add a small amount of water, and bring it to a boil. Plunge in shredded pork, remove as soon as pork shreds are free and moving around in water. Save the boiled water in a bowl for later use.

   In the cleaned wok, heat vegetable oil up, stir fry the blanched pork shreds until aromatic, add the chopped spring onions and ginger, cooking wine, and soy sauce. Stir in the noodles, add the boiled water, season with salt, continue to stir-fry. Add spinach, scrambled eggs when the soup is reduced, stir evenly. Remove all the ingredients from the wok and place in a bowl. Serve.

SANXIAN DUMPLINGS

Ingredients
Strong flour 500g
Minced pork 250g
Prawns 80g
Sea cucumber 80g (optional)
Eggs 4

Seasonings
Soy sauce 50g
Cooking wine 30g
Sesame oil 30g
Minced ginger 15g
Minced spring onions 15g
Salt 8g

Method

1. Forming the dough
   Place the flour on the worktable surface and make a hole in the middle with your hand, add some of the hot water into the hole. Slowly incorporate the surrounding flour with the water reaching a stage that is called "ears of the wheat" which is when the flour comes together with the water to form a breadcrumb stage. This is when you can knead until a dough is formed. Be careful not to add too much water as you will flood the dough. Continue to form the dough, adding more water if necessary, constantly feel the dough making sure that it is not too dry or firm. Once the dough has formed, continue to develop the gluten in the dough by kneading until the surface is smooth. Leave it to rest for 15 minutes with a damp cloth over the top of the dough to prevent it from dry and forming a crust on the surface.

2. Preparing the filling
   Place the pork mince into a suitable seized bowl and add the cooking wine, soy sauce, salt and spring onions and begin to mix. Once incorporated, add the sesame oil, mix again and leave it in the bowl for later use.
   Take the sea cucumber, cut lengthways and then dice into small cubes, meanwhile slice the prawns horizontally into small pieces. Place all on top of the pork mince including the four eggs scrambled.

3. Forming the dumplings
   Take the rested dough and split a third of the dough off lengthways, then roll it into a long cylinder. Take the dough in one hand and use your thumb and fore finger of the other hand to snap the dough stick into small pieces about 20g each. Once you have completed, scatter these pieces in flour to prevent them from sticking together.
   Take a piece of dough, flatten it, and roll it out with a small rolling pin into a circle. Spin the dough around and concentrate on working the edges of the dough to make them thinner than the centre. This will prevent the dumpling from being too thick in the areas it is joined together when sealed.
   Once you have enough of the circular dumpling doughs rolled out, hold one using your non dexter-

ous hand and begin to fill with the pork mince, prawns, sea cucumber and scrambled egg. Then fold in half and crimp with both of your thumbs and fore fingers.

4. Cooking

   Boil a pan of water and begin to add the dumplings one by one, only add a suitable number of dumplings to the water at once to allow enough space to stir them in the water. Keep the dumplings moving in a circular motion and once they float to the surface after 5 to 7 minutes, they are ready to be removed from the water and served.

Part Three

LAMB DUMPLINGS

Ingredients
Strong flour 500g
Water 280ml
Minced lamb 250g
Courgette 200g

Seasonings
Soy sauce 50g
Cooking wine 30ml
Szechuan pepper infused water 20g
Salt 8g
Minced ginger 15g
Minced spring onions 15g
Sesame oil 30g

Method

1. Forming the dough
   Place the flour on the worktable surface and make a hole in the middle with your hand, add some of the hot water into the hole. Slowly incorporate the surrounding flour with the water reaching a stage that is called "ears of the wheat" which is when the flour comes together with the water to form a breadcrumb stage. This is when you can knead until a dough is formed. Be careful not to add too much water as you will flood the dough. Continue to form the dough, adding more water if necessary, constantly feel the dough making sure that is not too dry or firm. Once the dough has formed, continue to develop the gluten in the dough by kneading until the surface is smooth. Leave to rest with a damp cloth over the top of the dough to prevent it from dry and forming a crust on the surface.

2. Preparing the filling
   Place the lamb mince into a bowl and add the cooking wine, 30ml Sichuan water (2g of Sichuan peppercorns in 200ml of hot water left to cool and use), soy sauce, 2g of salt and spring onions and begin to mix. Once incorporated, add the sesame oil, mix again, and leave the filling in the bowl for later use.
   Take the courgette and shred it into a bowl, season with 6g of salt and leave to one side to extract the water. Squeeze the excess water out of the courgette and place into the lamb mixture and combine.

3. Forming the dumplings
   Take the rested dough and split a third of the dough off lengthways, then roll it into a long cylinder. Take the dough in one hand and use your thumb and fore finger of the other hand to snap the dough stick into small pieces about 20g each. Once you have completed, scatter these pieces in flour to prevent them from sticking together.
   Take a piece of dough, flatten it, and roll it out with a small rolling pin into a circle. Spin the dough

around and concentrate on working the edges of the dough to make them thinner than the centre. This will prevent the dumpling from being too thick in the areas it is joined together when sealed. Once you have enough of the circular dumpling doughs rolled out, hold one using your non dexterous hand and begin to fill with the pork mince, prawns, sea cucumber and scrambled egg.
Then fold in half and crimp with both of your thumbs and fore fingers.

4. Cooking
Boil a pan of water and begin to add the dumplings one by one, only add a suitable number of dumplings to the water at once to allow enough space to stir them in the water. Keep the dumplings moving in a circular motion and once they float to the surface after 5 to 7 minutes, they are ready to be removed from the water and served.

Part Three

VEGETABLE OIL FLATBREAD

Ingredients
Strong flour 500g
Water 300g
Vegetable oil 200g

Seasonings
Salt 20g

Method

1. Forming the dough
   Place the flour on the worktable surface and make a hole in the middle with your hand, add some of the water into the hole. Slowly incorporate the surrounding flour with the water reaching a stage that is called "ears of the wheat" which is when the flour comes together with the water to form a breadcrumb stage. This is when you can knead until a dough is formed. Be careful not to add too much water as you will flood the dough. Continue to form the dough, adding more water if necessary, constantly feel the dough making sure that it is not too dry or firm. Once the dough has formed, continue to develop the gluten in the dough by kneading until the surface is smooth. Leave it to rest with a damp cloth over the top of the dough to prevent it from dry and forming a crust on the surface.

2. Pulling the noodles
   Once the dough has rested for 30 minuets, continue to form the gluten in the dough by rolling the dough into a long cylindrical shape. Then pickup at each end of the dough and begin to spin the dough into a plat. Once the gluten has been developed to the correct consistency begin to pull the dough to form thick noodles.

3. Forming the flatbread
   Brush the surface of the noodles with vegetable oil in the direction of the noodles. Now twist the noodles into a one cylindrical piece of dough, keeping the shape of the noodles but in one roll. Now roll one end of the dough into spiral to form a dough ball and cut it off from the cylindrical dough and wrap the end underneath the dough ball, then place it on a wooden tray to be cooked later. With the remaining dough, form as many spiralised dough balls as you can, placing each one of them on the wooden tray.

4. Cooking
   Oil the flatbread cooker. When the correct temperature is reached, begin to place the dough balls on the flatbread maker and flatten them further with your hand. Once golden brown on one side turns them over and repeat the process, close the lid and cook until the dough is completely cooked through. Pad away the excess oil with a kitchen towel and serve.

Chinese Culinary Techniques

SPRING FLATBREAD

Ingredients
Strong flour 500g
Hot water 200g
Colder water 50g
Vegetable oil 100g

Method

1. Forming the dough
   Place the flour on the worktable surface and make a hole in the middle with your hand, add some of the water into the hole. Slowly incorporate the surrounding flour with the water reaching a stage that is called "ears of the wheat" which is when the flour comes together with the water to form a breadcrumb stage. This is when you can knead until a dough is formed. Be careful not to add too much water as you will flood the dough. Continue to form the dough, adding more water if necessary, constantly feel the dough making sure that it is not too dry or firm. Once the dough has formed, continue to develop the gluten in the dough by kneading until the surface is smooth. Leave it to rest with a damp cloth over the top of the dough to prevent it from dry and forming a crust on the surface.

2. Shaping breads
   Take the rested dough and split a third of the dough off length ways, then roll into a long cylinder. Take the dough in one hand and use your thumb and fore finger of the other hand to snap the dough stick into small pieces. Once you have completed, scatter these pieces in flour to prevent them from sticking together. Once you have completed scatter in flour and roll them to prevent them from sticking together. Take each piece of dough and flatten it to a disk, brush with vegetable oil and flour. Place two disks on top of each other and flatten with your hands again. Separate the doughs and use a rolling pin to roll each disk into a side plate sized disk.

3. Cooking
   Place each dough disk onto the flatbread cooker, once the dough is cooked and starts to show heat spots, turn it over and cook the other side.

Part Three

STEAMED DUMPLINGS

Ingredients
Strong flour 500g
Hot water 250g
Scrambled egg 100g
Courgette grated 200g
Dried shrimp 30g

Seasonings
Sesame oil 30g
Salt 8g
Minced ginger 15g
Minced spring onions 15g

Method

1. Forming the dough
   Place the flour on the worktable surface and make a hole in the middle with your hand, add some of the hot water into the hole. Slowly incorporate the surrounding flour with the water reaching a stage that is called "ears of the wheat" which is when the flour comes together with the water to form a breadcrumb stage. This is when you can knead until a dough is formed. Be careful not to add too much water as you will flood the dough. Continue to form the dough, adding more water if necessary, constantly feel the dough making sure that it is not too dry or firm. Once the dough has formed, continue to develop the gluten in the dough by kneading until the surface is smooth. Leave it to rest with a damp cloth over the top of the dough to prevent it from dry and forming a crust on the surface.

2. Preparing the filling
   Grate the courgette into a bowl and season. The salt will encourage the excess water to be drawn from the grated courgette, take a handful and squeeze the excess water out. Add the scrambled egg, dried shrimp, salt the sesame oil and fold through.

3. Forming the dumplings
   Take the rested dough and split a third of the dough off lengthways, then roll it into a long cylinder. Take the dough in one hand and use your thumb and fore finger of the other hand to snap the dough stick into small pieces about 20g each. Once you have completed, scatter these pieces in flour to prevent them from sticking together.
   Take a piece of dough, flatten it, and roll it out with a small rolling pin into a circle. Spin the dough around and concentrate on working the edges of the dough to make them thinner than the centre. This will prevent the dumpling from being too thick in the areas it is joined together when sealed.
   Once you have enough of the circular dumpling doughs rolled out, hold one using your non dexterous hand and begin to fill with the pork mince, prawns, sea cucumber and scrambled egg.
   Then fold in half and crimp with both of your thumbs and fore fingers.

4. Cooking
   Once completed, place the dumplings in a steamer basket, ensure that there is enough space between each dumpling to cook evenly. Place the stacked steamer baskets into boiling water and begin to steam for around 10 minutes.

Part Three

FRIED DUMPLINGS

Ingredients
Strong flour 500g
Hot water 250g
Minced pork 250g
Prawns 80g
Sea cucumber (optional) 80g
Eggs 4
Corn starch 20g
Vegetable oil 50g

Seasonings
Sesame oil 30g
Salt 2g
Minced ginger 15g
Minced spring onions 15g
Soy sauce 30g
Cooking wine 30g

Method

1. Forming the dough
   Place the flour on the worktable surface and make a hole in the middle with your hand, add some of the water into the hole. Slowly incorporate the surrounding flour with the water reaching a stage that is called "ears of the wheat" which is when the flour comes together with the water to form a breadcrumb stage. This is when you can knead until a dough is formed. Be careful not to add too much water as you will flood the dough. Continue to form the dough, adding more water if necessary, constantly feel the dough making sure that it is not too dry or firm. Once the dough has formed, continue to develop the gluten in the dough by kneading until the surface is smooth. Leave it to rest with a damp cloth over the top of the dough to prevent it from dry and forming a crust on the surface.

2. Preparing the filling
   In a mixing bowl, add the cooking wine, soy sauce, salt, spring onions and ginger to the minced pork and combine together until fully incorporated, then add the sesame oil. Combine again and leave to one side until required. Prepare the Sea cucumber by dicing it and then slice the prawns into small pieces. Scramble the eggs and leave all the ingredients alongside the prior prepared pork mince.
   Take the cornflour and water and mix until a smooth paste is formed add a small amount of vegetable oil and combine.

3. Forming the dumplings
   Take the rested dough and split a third of the dough off lengthways, then roll it into a long cylinder. Take the dough in one hand and use your thumb and fore finger of the other hand to snap the dough stick into small pieces about 20g each. Once you have completed, scatter these pieces in flour to prevent them from sticking together.
   Take each piece of dough and flatten it to a disk and use a rolling pin to roll each disk into a suitable

sized dumpling wrapper.

Once all the wrappers are prepared, take one and place the correct amount of pork mince filling into the centre along with the sea cucumber, scrambled egg, and prawns. Close by folding the disk in half and pinch a section of the top only to seal.

4. Cooking

Brush the surface of a flat bread cooker with vegetable oil and place the dumplings on top in rows of seven. Cover with the cornflour solution and close the cooker top and cook until the base is golden brown, and the dumpling is thoroughly cooked. Lift off the cooker top and serve.

SANXIAN SHAOMAI (SIU MEI)

Ingredients
Strong flour 500g
Hot water 250g
Pork mince 300g
Prawns 80g
Re-hydrated dried sea cucumber (optional) 80g
Scrambled eggs 4

Seasonings
Soy sauce 50g
Cooking wine 30g
Sesame oil 30g
Minced spring onions 15g
Minced ginger 15g
Salt 8g

Method

1. Forming the dough
   Place the flour on the worktable surface and make a hole in the middle with your hand, add some of the hot water into the hole. Slowly incorporate the surrounding flour with the water reaching a stage that is called "ears of the wheat" which is when the flour comes together with the water to form a breadcrumb stage. This is when you can knead until a dough is formed. Be careful not to add too much water as you will flood the dough. Continue to form the dough, adding more water if necessary, constantly feel the dough making sure that it is not too dry or firm. Once the dough has formed, continue to develop the gluten in the dough by kneading until the surface is smooth. Leave it to rest for 15 minutes with a damp cloth over the top of the dough to prevent it from dry and forming a crust on the surface.

2. Preparing the filling
   In a bowl, place the minced pork, spring onions, cooking wine, soy sauce, salt, and combine to a smooth paste, then add the spring onions and sesame oil and combine again. Leave it aside for use later.
   Prepare the sea cucumber into small dices and cut the prawns into small sections. Scramble the eggs and again leave alongside with the pork mince for later use.

3. Forming the Shaomai
   Take the rested dough and split a third of the dough off lengthways, then roll it into a long cylinder. Take the dough in one hand and use your thumb and fore finger of the other hand to snap the dough stick into small pieces about 25g each. Once you have completed, scatter these pieces in flour and roll them to prevent them from sticking together. Take each piece of dough and flatten it to a disk and flour. Separate the doughs and use a rolling pin to roll each disk into a suitable sized disk.
   Once all the disks are prepared, take one disk and roll with a rolling pin that is thinner at one end. Place flour on the surface of the table and begin to roll the disk out further with the rolling pin by spinning the dough around by keeping the thinner part of the rolling still and work the wider part un-

til the dough is almost stretched and crimped on the circumference.

To form the dumpling fill the rolled-out disk with the pork mince, sea cucumber, scrambled egg, and prawns. Then take the filling and disk and push the filling down in pastry disk creating a pouch of the filling encased in pastry. Crimp the neck of the poach to create a collar and excess pastry with a little of the filling exposed. Tidy the top and place in a steaming basket.

4. Cooking

Brush the surface of a flat bread cooker with vegetable oil and place the Shaomai on top in rows of seven. Cover with the cornflour solution and close the cooker top and cook until the base is golden brown, and the Shaomai is thoroughly cooked. Lift off the cooker top and serve.

LAMB SHAOMAI (SIU MEI)

Ingredients
Strong flour 500g
Hot water 250g
Lamb mince 250g
Courgette 200g

Seasonings
Soy sauce 50g
Cooking wine 30g
Sesame oil 30g
Minced spring onion 10g
Minced ginger 10g
Salt 4g

Method

1. Forming the dough
   Place the flour on the worktable surface and make a hole in the middle with your hand, add some of the hot water into the hole. Slowly incorporate the surrounding flour with the water reaching a stage that is called "ears of the wheat" which is when the flour comes together with the water to form a breadcrumb stage. This is when you can knead until a dough is formed. Be careful not to add too much water as you will flood the dough. Continue to form the dough, adding more water if necessary, constantly feel the dough making sure that it is not too dry or firm. Once the dough has formed, continue to develop the gluten in the dough by kneading until the surface is smooth. Leave it to rest for 15 minutes with a damp cloth over the top of the dough to prevent it from dry and forming a crust on the surface.

2. Preparing the filling
   Grate the courgette into a bowl and season with salt. The salt will encourage the excess water to be drawn from the grated courgette, take a handful and squeeze the excess water out.
   In a bowl, add the lamb mince, ginger, soy sauce, salt, infused Sichuan peppercorn water and combine to a smooth paste. Finally, add the spring onions and sesame oil and the courgette and combine once more.

3. Forming the Shaomai
   Take the rested dough and split a third of the dough off lengthways, then roll it into a long cylinder. Take the dough in one hand and use your thumb and fore finger of the other hand to snap the dough stick into small pieces about 25g each. Once you have completed, scatter these pieces in flour and roll them to prevent them from sticking together. Take each piece of dough and flatten it to a disk and flour. Separate the doughs and use a rolling pin to roll each disk into a suitable sized disk.
   Once all the disks are prepared, take one disk and roll with a rolling pin that is thinner at one end. Place flour on the surface of the table and begin to roll the disk out further with the rolling pin by spinning the dough around by keeping the thinner part of the rolling still and work the wider part un-

til the dough is almost stretched and crimped on the circumference.

To form the dumpling fill the rolled-out disk with the pork mince, sea cucumber, scrambled egg, and prawns. Then take the filling and disk and push the filling down in pastry disk creating a pouch of the filling encased in pastry. Crimp the neck of the poach to create a collar and excess pastry with a little of the filling exposed. Tidy the top and place in a steaming basket.

4. Cooking

Before cooking, ensure that there is enough space between each Shaomai to cook evenly. Place the stacked steamer baskets over boiling water and begin to steam for around 10 minutes.

Part Three

FOUR-HAPINESS DUMPLINGS

Ingredients
Strong flour 500g
Hot water 250g
Pork mince 200g
Chopped black ear mushroom 100g
Chopped carrot 100g
Chopped pak choi 50g
Scrambled eggs 100g
Skinless prawns 30g

Seasonings
Sesame oil 30g
Cooking wine 10g
Light soy sauce 10g
Minced spring onions 15g
Minced ginger 15g
Salt 8g

Method

1. Forming the dough
   Place the flour on the worktable surface and make a hole in the middle with your hand, add some of the hot water into the hole. Slowly incorporate the surrounding flour with the water reaching a stage that is called "ears of the wheat" which is when the flour comes together with the water to form a breadcrumb stage. This is when you can knead until a dough is formed. Be careful not to add too much water as you will flood the dough. Continue to form the dough, adding more water if necessary, constantly feel the dough making sure that it is not too dry or firm. Once the dough has formed, continue to develop the gluten in the dough by kneading until the surface is smooth. Leave it to rest for 15 minutes with a damp cloth over the top of the dough to prevent it from dry and forming a crust on the surface.

2. Preparing the dumpling filling
   In a mixing bowl, place the minced pork, ginger, soy sauce, salt, cooking wine, sesame oil and combine them together. Add the prawns and combine again, leave the filling to one side to use later.

3. Forming the dumplings
   Take the rested dough and split a third of the dough off lengthways, then roll it into a long cylinder. Take the dough in one hand and use your thumb and fore finger of the other hand to snap the dough stick into small pieces about 20g each. Once you have completed, scatter these pieces in flour to prevent them from sticking together.
   Take each piece of dough and flatten it to a disk and flour. flatten with your hand and using a rolling pin to roll each disk into a suitable size.
   Once all the disks are prepared, take one disk and place the pork filling in the centre. Fold the disk in half and pinch together at the top to form two compartments. Turn 90 degrees and pinch together again to create four compartments. Take each of the four compartments and pinch them together

where they touch each other. Form the compartments so they are square in appearance, fill with the scrambled eggs, black ear mushrooms, chopped pak choi and carrots then place in a streamer tray.

4. Cooking
   Before cooking ensure that there is enough space between each dumpling to cook evenly. Place the stacked steamer baskets into boiling water and begin to steam for around 10 minutes.

Part Three

STEAMED BUNS

Ingredients
Strong flour 1,000g
Yeast 10g
Baking powder 10g
Water 550ml

Method

1. Forming the dough
   Place the flour on the worktable surface, mix with the yeast and baking powder. Make a hole in the middle of the pile with your hand, gradually add the water into the hole. Slowly incorporate the surrounding flour with the water reaching a stage that is called "ears of the wheat" which is when the flour comes together with the water to form a breadcrumb stage. This is when you can knead until a dough is formed. Be careful not to add too much water as you will flood the dough. Continue to form the dough, adding more water if necessary, constantly feel the dough making sure that it is not too dry or firm. Once the dough has formed, continue to develop the gluten in the dough by kneading until the surface is smooth. Leave it to rise for 15 minutes with a damp cloth over the top of the dough to prevent it from dry and forming a crust on the surface.

2. Moulding the dough
   Take a section of the dough and separate it into dough pieces of 75g each, shape a piece into a tight ball and place smooth side up onto a steamer tray. Ensure there is enough space for each dough ball to rise. Let them prove for 25 minutes at 40°C prior to steaming.

3. Cooking
   Bring water to the boil in a steamer, steam the buns for 15 minutes until they have cooked thoroughly and are white in appearance and when you touch them they should appear to be light.

Chinese Culinary Techniques

STEAMED TWISTS

Ingredients
Strong flour 1,000g
Yeast 10g
Baking powder 10g
Water 550ml
Vegetable oil 50ml

Method

1. Forming the dough
   Place the flour on the worktable surface, mix with the yeast and baking powder. Make a hole in the middle of the pile with your hand, gradually add the water into the hole. Slowly incorporate the surrounding flour with the water reaching a stage that is called "ears of the wheat" which is when the flour comes together with the water to form a breadcrumb stage. This is when you can knead until a dough is formed. Be careful not to add too much water as you will flood the dough. Continue to form the dough, adding more water if necessary, constantly feel the dough making sure that it is not too dry or firm. Once the dough has formed, continue to develop the gluten in the dough by kneading until the surface is smooth. Leave it to rise for 15 minutes with a damp cloth over the top of the dough to prevent it from dry and forming a crust on the surface.

2. Moulding the dough
   Roll the dough out to a rectangle pastry sheet about 6mm thick and then brush with vegetable oil and sprinkle with flour. Roll the sheet up from the top to the bottom to form a cylindrical shape and cut into sections that are 3cm wide. Take each piece of dough and press it vertically with a chop stick down the centre. Fold on top of itself twice from the top to the bottom of the dough, finally indent vertically again with a chop stick to complete. Cover the rolled buns with a cloth and leave for 20 minutes to prove in a steamer tray ensuring to leave enough space to cook evenly.

3. Cooking
   Place the steamer baskets into boiling water and steam the buns for 15 minutes.

Part Three

THOUSAND-LAYERED FLATBREAD

Ingredients
Strong flour 1,000g
Water 550ml
Yeast 5g
Baking powder 5g
Vegetable oil 25ml
Five-spice powder 20g
Salt 2g

Method

1. Forming the dough
   Place the flour on the worktable surface, mix with the yeast and baking powder. Make a hole in the middle of the pile with your hand, gradually add the water into the hole. Slowly incorporate the surrounding flour with the water reaching a stage that is called "ears of the wheat" which is when the flour comes together with the water to form a breadcrumb stage. This is when you can knead until a dough is formed. Be careful not to add too much water as you will flood the dough. Continue to form the dough, adding more water if necessary, constantly feel the dough making sure that it is not too dry or firm. Once the dough has formed, continue to develop the gluten in the dough by kneading until the surface is smooth. Leave it to rise for 15 minutes with a damp cloth over the top of the dough to prevent it from dry and forming a crust on the surface.

2. Moulding the dough
   Take the rested dough and roll it into a long cylinder, take the dough in one hand and use your thumb and fore finger to separate small pieces of the dough about 50g each. Once you have completed, scatter in flour and roll them together to prevent them from sticking together. Take each piece of dough and flatten it to a disk and flour.
   flatten with your hand and using a rolling pin to roll each disk into a long rectangle and brush with vegetable oil and sprinkle with 5-spice powder and a little bit of flour. Fold over from the top to the bottom in sections that are 5cm wide, flatten the ends and fold underneath the bun and roll again to slightly increase in seize.
   Place on a steamer tray and prove for 20 minutes.

3. Cooking
   Place the steamer trays into the oven and cook for 15 minutes, once removed cut in half and serve.

MENN DING BUNS

Ingredients
Strong flour 1,000g
Water 550ml
Yeast 10g
Baking powder 10g
Red bean paste 500g

Method
1. Forming the dough
   Place the flour on the worktable surface, mix with the yeast and baking powder. Make a hole in the middle of the pile with your hand, gradually add the water into the hole. Slowly incorporate the surrounding flour with the water reaching a stage that is called "ears of the wheat" which is when the flour comes together with the water to form a breadcrumb stage. This is when you can knead until a dough is formed. Be careful not to add too much water as you will flood the dough. Continue to form the dough, adding more water if necessary, constantly feel the dough making sure that it is not too dry or firm. Once the dough has formed, continue to develop the gluten in the dough by kneading until the surface is smooth. Leave it to rise for 15 minutes with a damp cloth over the top of the dough to prevent it from dry and forming a crust on the surface.
2. Moulding the dough
   Take the rested dough and roll it into a long cylinder, take the dough in one hand and use your thumb and fore finger to separate small pieces of the dough about 40g each. Scatter them in flour and roll them to prevent them from sticking together. Form each piece into a ball and leave them to rest. Cut the bean curd into blocks of 30g. Take a ball of dough and roll it to a disk again and place the bean curd into the centre. Cover the bean curd with the dough and seal at the base, remove the excess dough and place seal down in a steamer tray.
3. Cooking
   Leave the buns to prove for 20 mins at 40°C then steam the buns for 15 minutes until they have cooked thoroughly and are white in appearance and if you touch them, they should appear to be light.

SLIVER-THREAD ROLLS

Ingredients
Strong flour 1,000g
Water 560ml
Yeast 10g
Baking powder 10g
Caster sugar 200g
Vegetable oil 250g
Sesame oil 10g

Method

1. Forming the dough
   Place the flour on the worktable surface, mix with the yeast and baking powder. Make a hole in the middle of the pile with your hand, gradually add the water into the hole. Slowly incorporate the surrounding flour with the water reaching a stage that is called "ears of the wheat" which is when the flour comes together with the water to form a breadcrumb stage. This is when you can knead until a dough is formed. Be careful not to add too much water as you will flood the dough. Continue to form the dough, adding more water if necessary, constantly feel the dough making sure that it is not too dry or firm. Once the dough has formed, continue to develop the gluten in the dough by kneading until the surface is smooth. Leave it to rise for 15 minutes with a damp cloth over the top of the dough to prevent it from dry and forming a crust on the surface.

2. Pulling the noodles
   Once the dough has rested, take a third of the dough and begin to roll it out into a cylindrical shape. Hold the dough end to end with both hands, use your thumb to create a hole in one end and hold the dough normally with the other hand. Begin extending the dough by throwing it up and down until it stretches. When the dough is at the correct length, interlock your hands to the dough, spin, and interlock to form an eye. Repeat this process until the gluten is developed correctly, place the dough on the table and twist each ends and cover on flour. Pull the dough to develop the noodles to the desired size and brush them with vegetable oil. Begin to roll the dough up into a thicker cylindrical roll, the vegetable oil will make sure that the strands of noodles do not stick together. Cut the dough into 10cm long sections and place to one side to use later.
   Take another piece of dough and form into another cylindrical shape, holding the dough in one hand use your thumb and forefinger on the other hand to separate the dough into pieces and flour them so they do not stick together. Roll them into little dough balls and then flatten them into flat disks. Now begin to roll them out into larger sized wrappers.
   Take a wrapper and place the segmented pulled noodles in the centre. Fold both ends of the wrapper over to cover the noodles. Shape it into a bun. Place with the seam down and leave it to prove for 20 minutes.

3. Cooking
   Place each proved roll into steamer baskets, leaving the correct space between each roll so them steam evenly. Steam for 15 minutes and remove.

Chinese Culinary Techniques

TIANJIN BAOZI

Ingredients
Strong flour 500g
Yeast 5g
Baking powder 5g
Water 350g
Pork mince 500g

Seasonings
Light soy sauce 30g
Dark soy sauce 30g
Cooking wine 20g
Sesame oil 30g
Minced spring onions 50g
Minced ginger 50g
Salt 8g

Method

1. Forming the dough
   Place the flour on the worktable surface, mix with the yeast and baking powder. Make a hole in the middle of the pile with your hand, gradually add the water into the hole. Slowly incorporate the surrounding flour with the water reaching a stage that is called "ears of the wheat" which is when the flour comes together with the water to form a breadcrumb stage. This is when you can knead until a dough is formed. Be careful not to add too much water as you will flood the dough. Continue to form the dough, adding more water if necessary, constantly feel the dough making sure that it is not too dry or firm. Once the dough has formed, continue to develop the gluten in the dough by kneading until the surface is smooth. Leave it to rise for 15 minutes with a damp cloth over the top of the dough to prevent it from dry and forming a crust on the surface.

2. Preparing the filling
   In a mixing bowl, place the pork mince, minced ginger, cooking wine, salt, soy sauce, and light soy sauce. Mix all the ingredients together until smooth. Leave the minced spring onions in sesame oil to infuse. Add some water into the filling, and then the sesame oil with spring onions and mix again until fully incorporated. Leave to one side until required for forming the Baozi.

3. Shaping the Baozi
   Take the rested dough and split a third of the dough off lengthways, then roll into a long cylinder. Take the dough in one hand and use your thumb and fore finger to split up the cylinder into small dough pieces of 25 each. Once you have completed, scatter in flour and roll them together to prevent them from sticking together. Take each piece of dough and flatten it to a disk and roll it out to a larger wrapper. Once all the wrappers are prepared, take one wrapper in your left hand and place the correct amount of filling, enclose the dough disk around the filling. You can begin to start crimping with your right thumb and forefinger in a vertical position and turn the dough with your other hand. Only crimp with your fore finger and on the side of the dumpling that facing away from you by pull-

ing the dough in an upwards motion. Once completed ensure the crimps are complete and showing no signs of the fillings leaking out.
4. Cooking
Once completed place in a steamer shelf, ensure that there is enough space between each Baozi to cook evenly. Leave aside to prove for a while and then place the stacked steamer baskets into boiling water and begin to steam for around 6 minutes until they are fully cooked.

Chinese Culinary Techniques

TIANJIN PICKLED VEGETABLE BAOZI

Ingredients
Strong flour 500g
Yeast 5g
Baking powder 5g
Water 300g
Pork mince 500g

Seasonings
Pickled leaf mustard 300g
Bamboo shoots 50g
Shiitake mushrooms 30g
Pickled mustard green 30g
Pixian board bean paste 30g
Soy sauce 30g
Sugar 20g
Corn starch solution 30g
Cooking wine 30g
Minced spring onions 20g
Minced ginger 20g
Salt 8g

Method

1. Forming the dough
   Place the flour on the worktable surface, mix with the yeast and baking powder. Make a hole in the middle of the pile with your hand, gradually add the water into the hole. Slowly incorporate the surrounding flour with the water reaching a stage that is called "ears of the wheat" which is when the flour comes together with the water to form a breadcrumb stage. This is when you can knead until a dough is formed. Be careful not to add too much water as you will flood the dough. Continue to form the dough, adding more water if necessary, constantly feel the dough making sure that it is not too dry or firm. Once the dough has formed, continue to develop the gluten in the dough by kneading until the surface is smooth. Leave it to rise for 15 minutes with a damp cloth over the top of the dough to prevent it from dry and forming a crust on the surface.

2. Preparing the filling
   Begin dicing the Shiitake mushrooms, bamboo shoots, pickled mustard greens, spring onions, and ginger. In a wok, add a ladle of vegetable oil and begin to cook off the spring onions, ginger and the Pixian board bean Paste until aromatic. Add the pork mince and pickled leaf mustard, fry together. Plunge in the mushrooms and bamboo shoots and continue to cook. Now add the cooking wine, soy sauce and some water and continue to cook. Add the sugar, salt, pickled mustard greens and finally, the cornflour solution. Cook for three minutes until it thickens. Place on the side for completing the Baozi once cooled.

3. Shaping the Baozi

Take the rested dough and split a third of the dough off lengthways, then roll into a long cylinder. Take the dough in one hand and use your thumb and fore finger to split up the cylinder into small dough pieces of 25 each. Once you have completed, scatter in flour and roll them together to prevent them from sticking together. Take each piece of dough and flatten it to a disk and roll it out to a larger wrapper. Once all the wrappers are prepared, take one wrapper in your left hand and place the correct amount of filling, enclose the dough disk around the filling. You can begin to start crimping with your right thumb and forefinger in a vertical position and turn the dough with your other hand. Only crimp with your fore finger and on the side of the dumpling that facing away from you by pulling the dough in an upwards motion. Once completed ensure the crimps are complete and showing no signs of the fillings leaking out.

4. Cooking

Once completed place in a steamer basket, ensure that there is enough space between each Baozi to cook evenly. Leave aside to prove for a while and then place the stacked steamer baskets into boiling water and begin to steam for around 6 minutes until they are fully cooked.

Chinese Culinary Techniques

TIANJIN CABBAGE BAOZI

Ingredients
Strong flour 500g
Water 280ml
Yeast 5g
Baking powder 5g
Heart cabbage 500g
Dried shrimps 100g

Seasonings
Minced ginger 20g
Minced spring onions 20g
Sesame oil 30g
Salt 6g

Method

1. Forming the dough
   Place the flour on the worktable surface, mix with the yeast and baking powder. Make a hole in the middle of the pile with your hand, gradually add the water into the hole. Slowly incorporate the surrounding flour with the water reaching a stage that is called "ears of the wheat" which is when the flour comes together with the water to form a breadcrumb stage. This is when you can knead until a dough is formed. Be careful not to add too much water as you will flood the dough. Continue to form the dough, adding more water if necessary, constantly feel the dough making sure that it is not too dry or firm. Once the dough has formed, continue to develop the gluten in the dough by kneading until the surface is smooth. Leave it to rise for 15 minutes with a damp cloth over the top of the dough to prevent it from dry and forming a crust on the surface.

2. Preparing the filling
   Shred the cabbage and blanch it in a wok with boiling water, refresh straight away in cold water and drain. In the wok, heat vegetable oil and add the ginger and spring onions then the dried prawns. Begin to add the cabbage but firstly make sure you have squeezed all the water out of the cabbage. Continue to cook and add the salt and sesame oil, empty onto a plate for use later.

3. Shaping the Baozi
   Take the rested dough and split a third of the dough off lengthways, then roll into a long cylinder. Take the dough in one hand and use your thumb and fore finger to split up the cylinder into small dough pieces of 25 each. Once you have completed, scatter in flour and roll them together to prevent them from sticking together. Take each piece of dough and flatten it to a disk and roll it out to a larger wrapper. Once all the wrappers are prepared, take one wrapper in your left hand and place the correct amount of filling, enclose the dough disk around the filling. You can begin to start crimping with your right thumb and forefinger in a vertical position and turn the dough with your other hand. Only crimp with your fore finger and on the side of the dumpling that facing away from you by pulling the dough in an upwards motion. Once completed ensure the crimps are complete and showing

no signs of the fillings leaking out.
4. Cooking
   Once completed place in a steamer basket, ensure that there is enough space between each Baozi to cook evenly. Leave aside to prove for a while and then place the stacked steamer baskets into boiling water and begin to steam for around 6 minutes until they are fully cooked.

Chinese Culinary Techniques

TIANJIN VEGETARIAN BAOZI

Ingredients
Strong flour 500g
Water 280ml
Yeast 5g
Baking powder 5g
Bean sprouts 200g
Fenpi (thick rice noodle) 250g
Five-spice dry tofu 50g
Gluten 50g
Shiitake mushrooms 15g
Dried yellow day lily flower 20g

Bean curd skin 30g
Black wood ear fungus 40g
White wood ear fungus 30g
Coriander 25g

Seasonings
Minced ginger 20g
Minced spring onion 20g
Sesame oil 30g
Salt 6g

Method

1. Forming the dough
   Place the flour on the worktable surface, mix with the yeast and baking powder. Make a hole in the middle of the pile with your hand, gradually add the water into the hole. Slowly incorporate the surrounding flour with the water reaching a stage that is called "ears of the wheat" which is when the flour comes together with the water to form a breadcrumb stage. This is when you can knead until a dough is formed. Be careful not to add too much water as you will flood the dough. Continue to form the dough, adding more water if necessary, constantly feel the dough making sure that it is not too dry or firm. Once the dough has formed, continue to develop the gluten in the dough by kneading until the surface is smooth. Leave it to rise for 15 minutes with a damp cloth over the top of the dough to prevent it from dry and forming a crust on the surface.

2. Preparing the filling
   Begin by dicing the five-spice tofu, the gluten and the Fenpi. Dice all the Shiitake mushrooms, dried yellow day lily flower, bean curd skin, black wood ear fungus, and white wood ear fungus. Now chop the coriander and then finely dice the ginger and spring onions. In your wok, begin to boil water and blanch the bean sprouts, refresh them in cold water immediately and drain, then chop them into small sections around 1cm in length. In the wok, heat star anise with the vegetable oil and add the bean curd and cook out. Add all the prior diced items into the wok, squeeze the bean sprout to remove the excess water and add to the wok. Continue to cook and add the sesame paste, salt, fermented flour paste, and the sesame oil. Cook until each item is cooked fully, then dish out and leave it aside until forming the Baozi.

Part Three

3. Shaping the Baozi
   Take the rested dough and split a third of the dough off lengthways, then roll into a long cylinder. Take the dough in one hand and use your thumb and fore finger to split up the cylinder into small dough pieces of 25 each. Once you have completed, scatter in flour and roll them together to prevent them from sticking together. Take each piece of dough and flatten it to a disk and roll it out to a larger wrapper. Once all the wrappers are prepared, take one wrapper in your left hand and place the correct amount of filling, enclose the dough disk around the filling. You can begin to start crimping with your right thumb and forefinger in a vertical position and turn the dough with your other hand. Only crimp with your fore finger and on the side of the dumpling that facing away from you by pulling the dough in an upwards motion. Once completed ensure the crimps are complete and showing no signs of the fillings leaking out.
4. Cooking
   Once completed place in a steamer basket, ensure that there is enough space between each Baozi to cook evenly. Leave aside to prove for a while and then place the stacked steamer baskets into boiling water and begin to steam for around 6 minutes until they are fully cooked.

MEAT ROLL

Ingredients
Strong flour 500g
Water 280ml
Yeast 5g
Baking powder 5g
Pork mince 200g

Seasonings
Chicken stock powder 5g
Light soy sauce 10g
Cooking wine 10g
Minced ginger 20g
Minced spring onions 20g
Sesame oil 10g
Salt 5g

Method

1. Forming the dough
   Place the flour, yeast and baking powder on the worktable surface and make a hole in the middle of the pile with your hand, gradually add the water into the hole. Slowly incorporate the surrounding flour with the water reaching a stage that is called "ears of the wheat" which is when the flour comes together with the water to form a breadcrumb stage. This is when you can knead until a dough is formed. Be careful not to add too much water as you will flood the dough. Continue to form the dough, adding more water if necessary, constantly feel the dough making sure that it is not too dry or firm. Once the dough has formed, continue to develop the gluten in the dough by kneading until the surface is smooth. Leave it to rise for 15 minutes with a damp cloth over the top of the dough to prevent it from dry and forming a crust on the surface.

2. Preparing the filling
   Add the pork mince into a bowl, add the ginger, soy sauce and cooking wine. Incorporate together and add the salt, chicken stock, spring onions and sesame oil. leave it aside until forming the meat roll.

3. Shaping the meat roll
   Use the rested dough to form a large rectangular sheet about 1cm thick, place the pork mince in the centre of the sheet, then spread it over leaving an inch of dough around all the sides. Roll the sheet from the top to the bottom keeping it in a cylindrical shape, then flatten each end of the roll with a rolling pin and fold underneath.

4. Cooking
   Place onto an oiled steamer tray and place and prove for 20 minutes. Then place into the steamer for 15 minutes, once cooked, remove from the steamer. Slice the cooked roll every inch to serve.

SESAME SHAOBING

Ingredients
Strong flour 1,000g
Yeast 10g
Baking powder 10g
Water 300g
Baking powder 10g
Vegetable oil 250g
White sesame seeds 200g

Seasonings
Five-spice powder 25g
Soy sauce 30g

Method

1. Forming the dough
   To make the dry oil shortening solution, heat up 250g of vegetable oil in a saucepan on medium heat. In a mixing bowl, mix 500g of flour and five-spice powder. Pour 250g of the heated vegetable oil gradually into the dry ingredients and incorporate until smooth.
   To make the skin dough, combine 500g of flour, 10g of yeast, 10g of baking powder, and 250g of water together to form a dough. Slowly incorporate the surrounding flour with the water reaching a stage that is called "ears of the wheat" which is when the flour comes together with the water to form a breadcrumb stage. This is when you can knead until a dough is formed. Once the dough has formed, continue to develop the gluten in the dough by kneading until the surface is smooth. Leave it to rest for 5 minutes under a cloth so the surface of the dough does not dry out.

2. Shaping the Shaobing
   Take the rested dough and roll it out into a large rectangle sheet, roughly 2mm thick. Pour the soy sauce into the dry oil shortening solution, spread it evenly onto the surface of the sheet, sprinkle with a little bit dry flour, and begin to roll the sheet from the top to the bottom. Use your thumb and fore finger to snap the roll into small pieces, shape them into small balls, slightly flatten them by pressing with your hand, then brush with water, finally dip each shaobing into sesame seeds.

3. Cooking
   Place each Shaobing onto a tray with the sesame seed side down ensuring enough space to cook evenly. Bake them at 180°C in a preheated the oven for 15 minutes. Flip side when the top side becomes golden.

Section 5 Chinese Soup Recipes

CHICKEN AND SWEET CORN SOUP

Ingredients
Sweet corn 400g
Chicken breast 100g
Egg white 50g

Seasonings
Salt 6g
Clear chicken broth 1,200g
Cornflour 5g
Cornflour solution 30g

Method
1. Cut the chicken breasts into small dices, mix with 1g of salt, 5g of cornflour, and 10g of water, leave it to marinate for 5 minutes.
2. Whisk the egg whites to achieve an even texture.
3. Blitz the sweet corn in a blender for 5 seconds. The texture should be grainy.
4. Place the chicken broth in wok, add the sweet corn. Bring it to the boil, season with 5g of salt, stir the chicken dices in, bring it to the boil.
5. Add cornflour solution in the soup, remove the wok from heat briefly to stir in egg whites.
6. Return wok back on stove, stir the soup.
7. Dish out and serve.

Features of the dish
This soup is thick and umami with a strong corn aroma.

Part Three

WEST LAKE BEEF SOUP

Ingredients
Beef tenderloin 150g
Soft tofu 50g
Straw mushrooms 100g
Egg whites 50g
Chopped coriander 5g

Seasonings
Cooking wine 50g
Light soy sauce 10g
White pepper powder 15g
Salt 6g
Clear beef broth 1,200g
Corn starch 5g
Corn starch solution 30g
Sesame oil 5g

Method
1. Soak the beef tenderloin in cold water to wash off blood stains, pad dry with kitchen paper towels, mince the beef manually, season it with cooking wine, light soy sauce, white pepper powder 5g, salt 1g, corn starch 5g, stir the beef in one direction to mix it well with seasonings. Set it aside to marinate for 10 minutes meanwhile mince the soft tofu and straw mushrooms.
2. Heat up water in a pot to almost boiling, lower the marinated beef in and keep stirring, add the minced tofu and strew mushrooms. Blanching all food items briefly and remove from the wok to drain.
3. Return the empty wok back to heat, pour the clear beef broth in, season with pepper powder and 5g of salt. Bring the soup to the boil, drop in the minced beef, straw mushroom, tofu, and corn starch solution. Quickly add whisked egg whites, add chopped coriander and sesame oil to the pot. Keep stirring.
4. Dish out and serve.

Features of the dish
The soup is savoury, pure and rich in flavour.

Chinese Culinary Techniques

SLIVER EAR MUSHROOM AND ASIAN PEAR SOUP

Ingredients
Rehydrated sliver ear mushrooms 150g
Asian pears 200g
Hawthorn berry jelly (fruit bar) 60g

Seasonings
Rock sugar 100g
Salt 1g
Corn starch solution 10g

Method
1. Break the white ear mushrooms into small pieces and place them in a bowl, add in 500g of water. Steam the bowl for about 1 hour until the mushrooms get jellified and sticky. Depending on the quality of the mushrooms, the cooking time may vary.
2. Peel the pears and cut them into 1.5x1.5x1.5cm dices. Cut the hawthorn berry jelly into 1x1x1cm dices.
3. Place 500g of water in a wok, plunge in the steamed mushrooms, bring it to a boil, simmer on low heat until the soup becomes thick and sticky, pour in diced pear, continue with the simmering for another 15 minutes, add in sock sugar and salt, cook for another 5 minutes.
4. Dish out, garnish with the hawthorn berry jelly dices, and serve.

Features of the dish
The soup is sweet, fruity, smooth, and thick with a hint of sourness.

BEEF AND TOMATO SOUP

Ingredients
Tomatoes 100g
Beef 80g
Cooked sweet peas 50g
Chopped onion 15g

Seasonings
Cooking wine 30g
Black pepper powder 2g
Salt 6g
Clear beef broth 1,200g
Cornflour 30g
Spring onions 25g
Cooking oil 2g
Sesame oil 3g

Method
1. Finely chop the beef, season with black pepper, cornflour, cooking wine, and salt 1g, leave it to marinate for 5 minutes.
2. Blanch the beef with boiling water, remove and set it aside.
3. Scald the tomatoes in boiling water to remove the skins and then cut them into 1x1x1cm dices.
4. Heat up a small amount of oil in wok, stir fry chopped onion until it becomes aromatic, stir in the tomato, 5g of salt, and sweet peas. Add the clear beef broth to the wok, plunge in beef, cook for a minute. Add the cornflour to thicken the soup.
5. Dish out, add a splash of sesame oil to the soup, garnish with the chopped spring onions and serve.

Features of the Dish
This soup is savoury, umami, slightly sour.

SOUR AND SPICY SOUP

Ingredients
Tofu 100g
Lean pork shreds 100g
Rehydrated dry black ear mushroom shreds 50g
Cooked bamboo shoots 50g
Spring onions shreds 10g
Beaten eggs 30g
Coriander mince 10g

Seasonings
White pepper powder 15g
Light soy sauce 10g
Fragrant vinegar 60g
Sesame oil 2g
Salt 4g
Clear chicken broth 1,000g
Corn starch solution 30g

Method
1. Shred the tofu and blanch it. Blanch the pork, bamboo shoots, and black ear mushrooms one by one with boiling water. Place all food items in a wok, add clear chicken broth, salt, and light soy sauce. Bring it to the boil, thicken with cornflour solution, reduce to low heat, stir in the beaten eggs. Increase to high heat when the egg sets. Remove the wok from heat when the shredded pork floats on the soup surface.
2. Dish out to a serving bowl, season with the white pepper powder, fragrant vinegar, spring onions, and sesame oil.
3. Serve.

Features of the Dish
It is sour, spicy, umami, and fragrant.

CHICKEN, CUCUMBER, AND EGG SOUP

Ingredients
Egg omelette 15g
Cucumber 50g
Chicken breast 100g
Clear chicken broth 1,200g
Egg white 3g
Rehydrated black ear mushrooms 10g
Cornflour 5g

Seasonings
Salt 6g
Olive oil 2g

method

1. Slice the chicken breast into 4cm long, 3cm wide, 0.3cm thick slices. Season with cornflour, egg white, salt 1g, mix well. Peel the cucumber and cut it into willow leave shape pieces, cut the egg omelette into 5cm long, 0.2cm thick strips, shred the mushrooms.
2. Blanch the chicken and mushroom with boiling water separately.
3. Pour the clear chicken broth into a wok, add salt, bring it to the boil, add the chicken slices, mushrooms, cucumber, and omelette strips. Mix well and remove the soup from heat.
4. Dish out, sprinkle the olive oil on top and serve.

Features of the Dish
This soup is savoury, umami, refreshing, and fragrant.

SCALLOP AND WINTER MELON SOUP

Ingredients
Scallop meat 200g
Peeled winter melon (wax gourd) 300g
Rehydrated glass (cellophane) noodles 150g

Seasonings
Salt 6g
Clear chicken broth 800g
Olive oil 2g

Method
1. Cut the winter melon into 2cm long, 0.3cm thick elephant-eye shape slices. Rinse the scallop.
2. Pour the clear chicken broth in a wok, add in the winter melon slices and bring it to the boil.
3. Plunge in the scallop, glass noodles, bring the soup to the boil again, season with salt and olive oil.
4. Dish out and serve.

Features of the Dish
The soup is umami, fragrant, and refreshing.

LOTUS SEED AND RED BEAN SOUP

Ingredients
Pre-cooked lotus seeds 200g
Red bean paste 400g
Cornflour 30g
Water 1,000g

Seasonings
Salt 1g
Sugar 150g

method
1. Make red bean puree by mixing red bean paste with water.
2. Place 1,000g of water in a wok, add sugar and salt to it. Bring it to the boil, add the red bean puree, bring the soup to the boil again, continue to cook for 2 minutes.
3. Thicken the soup by adding the cornflour.
4. Dish out and serve.

Features of the Dish
This soup is sweet, fragrant, and smooth.

PRAWNS AND TOFU SOUP

Ingredients
Fresh prawn meat 100g
Soft tofu 500g
Spring onion 5g
Fresh sweet peas 15g
Cornflour 50g
Clear chicken broth 1,000g

Seasonings
Salt 6g
White pepper powder 1g

Cooking Process
1. Cut the tofu into 1x1x1cm dices, chop the spring onions. Blanch the tofu, sweet peas separately with boiling water and leave them aside to drain.
2. Place the clear chicken soup in a wok, bring it to the boil. Plunge in tofu, prawns, sweet peas, salt, and white pepper powder.
3. Thicken the soup by adding the cornflour, mix them well.
4. Sprinkle the chopped spring onions and olive oil.
5. Dish out and serve.

Features of the Dish
The soup is savoury and umami. Tofu is soft and smooth.

FERMENTED RICE SOUP WITH GLUTINOUS RICE BALLS

Ingredients
Sweet fermented rice 400g
Glutinous rice flour 200g
Sugar-marinated osmanthus 30g
Lotus root powder 30g

Seasonings
White caster sugar 50g
Salt 1g

Method
1. Gradually add a small amount of water into the glutinous rice flour to make a thick paste. Take a small piece of the paste and roll it into a ball with both of your hands. Repeat the step until the paste is used up.
2. Bring 1,000g of water to the boil in a wok, plunge the sticky rice balls in, remove from the boiling water when balls float on the surface.
3. Add 600g of water in an empty work, season with sugar, sweet fermented rice, salt, and sugar-marinated Osmanthus. Bring it to the boil, thicken with lotus root powder solution.
4. Add in sticky rice balls.
5. Dish out and serve.

Features of the Dish
The glutinous rice balls are smooth, chewy, and soft. The soup has an appetising fermented rice aroma. .

SONGSAO FISH SOUP (AUNTY SONG'S FISH SOUP)

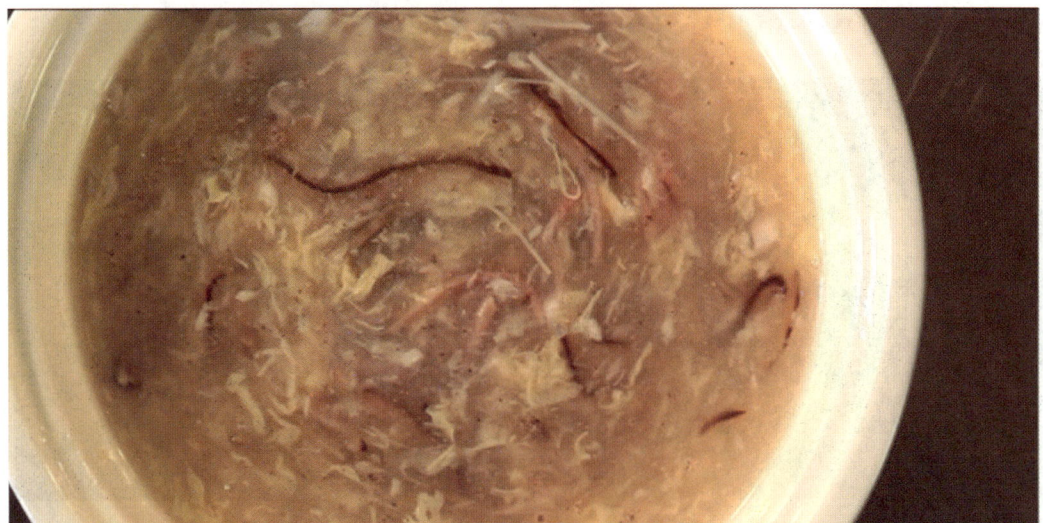

Ingredients
1 whole mandarin fish (about 600g)
Cooked dry cured ham (shredded) 10g
Cooked bamboo shoots (shredded) 25g
Rehydrated shiitake mushrooms (shredded) 25g

Seasonings
Beaten egg yolks 3
Spring onions segments 25g
Ginger (crushed) 5g
Shredded ginger 1g
White pepper powder 1g
Cooking wine 30g
Vegetable oil 25g
Salt 5g
Rice vinegar 25g
Clear broth 250g
Cornflour 30g

Method
1. Fillet the mandarin fish.
2. Leave the fish fillets on a plate with the skin side down, marinate them with the cooking wine, salt 1g, spring onions 10g, and ginger.
3. Place the plate in a pre-heated steamer and cook for 6 minutes on high heat. Turn the heat off.
4. Pick out the spring onions and ginger, pour the liquid into a deep plate. With a pair of chopsticks, remove the fish skin and bones, break the fish fillets into small pieces and leave them in the deep plate with the liquid.
5. Pre-heat the wok on high heat, add in 15g of oil, and stir fry the spring onions stems until aromatic, add the broth in, bring it to the boil, remove the spring onions.
6. Add 15g of cooking wine, shredded bamboo shoots, and shiitake mushrooms into the broth. Bring it to the boil again.
7. Plunge in the fish fillets, season with 4g of salt, slowly stir in cornflour solution, then add in the egg yolks, keep combining, season with the rice vinegar.
8. Place the soup in a soup bowl, sprinkle the shredded dry cured ham, shredded ginger, and white pepper on top. Stir and serve.

Features of the Dish
The soup is umami, tender and smooth with a taste akin to crab.

Part Three

Section 6 Chinese Dessert Recipes

LIU LI WALNUT (CARAMEL WALNUT)

Ingredients
Walnuts 300g
Vegetable oil 500g
Water 40g

Seasonings
Caster sugar 150g
Honey 30g

Method
1. Soak walnuts in hot water to soften the skin to remove it.
2. Blanch them in a wok with boiling water, refresh straight away in cold water and drain.
3. Heat 500g of vegetable oil in the wok to 120°C, fry the walnuts until they float on the oil surface and the colour darkened, remove from the hot oil.
4. Greasy a plate.
5. Add water, sugar, and honey in a saucepan, cook on high heat. Gently stir with a spatula to dissolve the sugar, reduce to low heat once the sugar completely dissolved, keep spreading the sugar syrup inside the saucepan.
6. When the sugar syrup starts bubbling and has turned to a thick and sticky caramel, plunge walnut in, swirl the pan and gently stir the walnut at the same time to make the coating even.
7. Place the caramel walnut on the oiled plate, separate them from each other.
8. Serve the coating hardens.

Chinese Culinary Techniques

GINGER CRIPSY DOUGH DRAIDS

Ingredients
Plain flour 500g
Corn starch 100g
Vegetable oil 290ml
Water 200ml
Caster sugar 200g
Sugared osmanthus fragrans 50g
dropped dried cranberries 50g
Ginger shreds 25

Method

1. Forming the dough
   Place the flour on the worktable surface and make a hole in the middle with your hand, gradually add the water into the hole. Slowly incorporate the surrounding flour with the water reaching a stage that is called "ears of the wheat" which is when the flour comes together with the water to form a breadcrumb stage. This is when you can knead until a dough is formed. Be careful not to add too much water as you will flood the dough. Continue to form the dough, adding more water if necessary, constantly feel the dough making sure that it is not to dry or firm. Once the dough has formed, continue to develop the gluten in the dough by kneading until the surface is smooth. Leave it to rest for 15 minutes with a damp cloth over the top of the dough to prevent it from dry and forming a crust on the surface.

2. Shaping the dough braid
   Once the dough has rested, roll it out into a large rectangle and dust with corn starch to prevent it from sticking together. To achieve this process, roll the dough out, then use the rolling pin to roll the dough up. Pull the rolling pin out of the dough and widen the dough by pressing it down repeatedly with the rolling pin. Repeat this step vertically and horizontally, then re-roll the dough out until the dough sheet becomes 1mm thick. Concertina the dough back and forth so the dough is folded up upon itself and is no wider than 6cm. Divide the dough stack into 3 equal sections.
   Take one section of the cut rolled dough and unroll it so it becomes single layer of dough stacked upon itself in long rectangles. Now begin to cut out into rectangular pieces around 6cm x 3cm. To each stack of dough sheets, fold it in half-lengthways and score three cuts no longer than 2cm. Open the stack to form a rectangle again, take the bottom of the rectangle and fold it from underneath through the centre to turn it inside out, to form a braid.

3. Cooking
   Once you have all your braids completed, cook off in a wok with hot vegetable oil, stir constantly to ensure they do not stick together. Once cooked, place them in a colander and drain the excess vegetable oil.

Part Three

To make the osmanthus sugar syrup, add three ladles of water and bring it to the boil. Add the caster sugar and sugared osmanthus fragrans in, bring it to the boil again, cook on high heat until it thickens to become syrup.

Mix the syrup with the braids in the colander and allow all the excess syrup to drain off and serve. Garnish with ginger shreds and the cranberries.

Chinese Culinary Techniques

SUGAR TRIANGLES

Ingredients
Plain flour 1,000g
Water 550ml
Yeast 10g
Baking powder 20g
Chinese brown sugar 200g

Method

1. Forming the dough
   Mix the flour, 10g of yeast, and 10g of baking powder on the worktable surface. Make a hole in the middle of the pile with your hand, gradually add the water into the hole. Slowly incorporate the surrounding flour with the water reaching a stage that is called "ears of the wheat" which is when the flour comes together with the water to form a breadcrumb stage. This is when you can knead until a dough is formed. Be careful not to add too much water as you will flood the dough. Continue to form the dough, adding more water if necessary, constantly feel the dough making sure that is not too dry or firm. Once the dough has formed, continue to develop the gluten in the dough by kneading until the surface is smooth. Leave it to rest for 15 minutes with a damp cloth over the top of the dough to prevent it from dry and forming a crust on the surface.
2. Moulding the bun
   Separate the dough into small pieces of 50g each and flatten them with your hand to from small wrappers.
   Add 10g of baking powder to the brown sugar and mix them together, take a wrapper, widen it with a rolling pin, place a teaspoon of brown sugar into the centre. Pinch the dough into three equal parts with your thumb and forefinger and ensure they sealed correctly and place in a steamer tray.
3. Cooking
   Leave the buns to prove for 20 mins at 40°C, then steam for 15 minutes until are fully cooked, when you touch them they should appear to be light.

STEAMED OSMANTHUS CAKE

Ingredients
Plain flour 1,000g
Water 620ml
Cornmill 250g
Yeast 10g
Baking powder 5g
Caster sugar 30g
Osmanthus fragrans jam 50g
Dried Fruit 50g

Method

1. Forming the dough
   Place the flour and cornmill on the worktable surface and make a hole in the middle with your hand, add the yeast, baking powder, osmanthus jam, and sugar into the hole and begin to add the water little by little. Slowly incorporate the surrounding flour with the water reaching a stage that is called "ears of the wheat" which is when the flour comes together with the water to form a breadcrumb stage. This is when you can knead until a dough is formed. Be careful not to add too much water as you will flood the dough. Continue to form the dough, adding more water if necessary, constantly feel the dough making sure that it is not to dry or firm. Once the dough has formed, continue to develop the gluten in the dough by kneading until the surface is smooth. Now roll the dough out so it will fit into the correct seized tray, brush water onto the surface and scatter the dried fruit on top.

2. Cooking
   Leave the dough to prove until it doubles in size, then place it into the steamer to steam over boiling water for 20 minutes. Remove from the steamer and cut into equally seized squares and serve.

GANLU SHORTCAKE

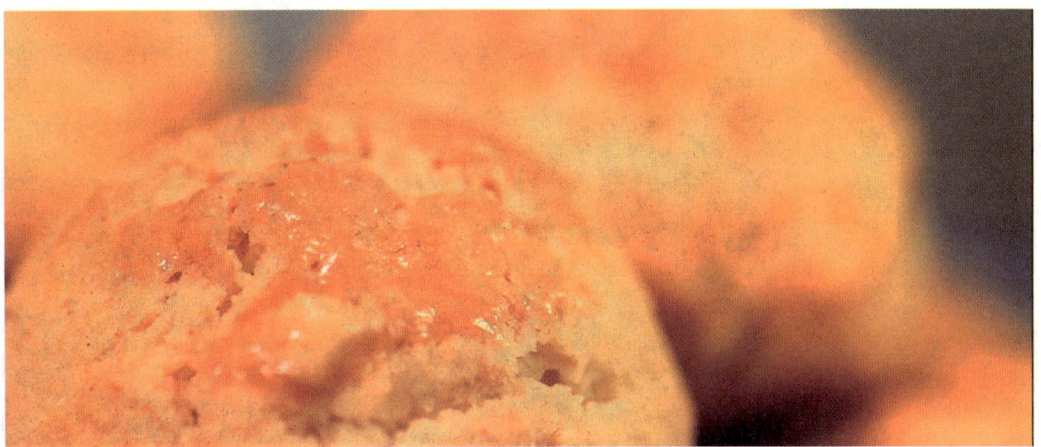

Ingredients
Flour 500g
Baking powder 10g
Ammonia powder 1.5g
Cheese powder 10g
Lard 150g
Butter 100g
Caster sugar 150g
Eggs 2
Red bean paste 500g

Method

1. Forming the dough
   Place the flour, cheese powder, sugar, ammonia powder, baking powder, lard and butter on the worktable surface and make a hole in the middle with your hand, add the eggs into the hole. Slowly incorporate the surrounding flour with the egg, butter and lard until reaching a stage that is called "ears of the wheat" which is when the flour comes together with the water to form a breadcrumb stage. This is when you can knead until a dough is formed. Once the dough has formed, continue to develop the dough kneading it until the surface is smooth.

2. Shaping the shortcake
   Roll the dough into a long strip, and then roll the red bean paste block into a long strip of the similar thickness. Separate a small piece (about 20g) from the dough, fill it with a small amount of red bean paste and form a ball. Place the formed dough ball on a baking tray.

3. Cooking
   Cover all the dough balls with egg wash and then bake at 180°C until golden brown then serve.

SAQIMA

Ingredients
Plain flour 500g
Ammonia powder 1.5g
Baking soda 2g
Egg 250g
Sugar 250g
Glucose 100g
Chopped dried fruit 50g
Sugar-marinated osmanthus 50g

Method
1. Forming the dough
   Mix the flour, sugar, ammonia powder, baking soda on the worktable surface and make a hole in the middle with your hand, add the eggs into the hole. Slowly incorporate the surrounding flour with the egg, butter, and lard until reaching a stage that is called "ears of the wheat" which is when the flour comes together with the water to form a breadcrumb stage. This is when you can knead until a dough is formed. Once the dough has formed, continue to develop the dough kneading it until the surface is smooth. Rest for 15 minutes with a damp cloth over the top of the dough to prevent it from dry and forming a crust on the surface.
2. Making the hand-cut noodles
   Roll the dough into a large rectangle and flour to prevent it from sticking to the worktable surface. To achieve this process, roll the dough out then use a rolling pin to roll the dough up. Pull the rolling pin out of the dough and widen the dough by pressing down on the dough repeatedly with the rolling pin. Repeat this vertically and horizontally, then re-roll the dough out to complete the required seize. For this recipe once the dough is rolled up on the rolling pin cut it lengthways along the rolling pin and place the opened-up dough on the chopping board. Fold it over and begin cutting it into thin strips about 0.5cm in thickness.
3. Cooking
   Fry the noodles in hot vegetable oil of 190°C until firm and then drain off the oil. In an empty wok add two ladles of water, sugar, glucose, and the sugar marinated osmanthus. Carry on cooking the sugar solution until it reduces and turns golden. Ladle the sugar solution over the noodles in a saucepan and bind together. Transfer into a frame and ensure it is evenly mixed. Sprinkle the mixed fruit over the top and then compress the noodle mixture and release from the frame. When it has firmed up and cooled, cut into squares to serve.

CAKE ROLLS

Ingredients
Egg 500g
White sugar 250g
Cake flour 250g
Water 100g
Vegetable oil 50g

For garnishing
Chocolate 100g
Fresh cut flowers 10
Dried fruit 50g
Single cream 500g

Method

1. Making the batter
   In a mixer add the eggs and sugar and whisk until white and aerated. Now add the flour and continue to whisk until light and fluffy and add the vegetable oil.
2. Baking the cake
   Pour the batter into a baking tray lined with grease proof paper and bake at 180°C for 30 minutes. Bake until a skewer come out clean from the centre of the cake, remove from the oven and allow to cool on a cooling wire and cut the cake in half.
3. Decorating the cake
   To decorate the cake, whisk the cream until soft peaks, take one half of the cake and cover the surface with whipped cream. Working from the bottom up take a rolling pin underneath the grease proof paper and roll the cake upwards to create a roll. Place the rolled sponge cake onto a plate to complete the decoration.
   Take a palette knife and apply the whipped cream all over the cake, ensure that the cream is applied evenly and there are no areas left uncovered. With a pipping bag use a star nozzle and being to pipe rosettes left to right alternatively down the centre of the cake roll.
   To garnish use the chocolate, fresh flowers, and dried fruit, arrange on the cake and then sprinkle with cocoa powder to finish.

YIPIN SHAOBING

Ingredients

For the dry oil shortening dough
Plain flour 500g
Lard 250

For the skin dough
Plain flour 500g
Lard 150g
Water 200g

For the filling
Roast flour (pre-roasted in the oven at 180°C until it turns golden) 200g
Sugar 400g
Sesame seeds 200g
Walnuts 100g
Peanuts 100g
Raisins 100g
Cranberry 100g
Dried winter melon strips 25g
Dried fruit 25g
Sesame oil 40g
Honey 40g
Maltose 100g

For bushing
Eaten egg yolks 50g

Method
1. Forming the dough
   To make the dry oil shortening, combine 250g of lard in 500g of plain flour together until a smooth dough has formed.
   To make the skin dough, combine 500g of plain flour, 150g of lard, and 200g of water together. Leave it to rest for 15 minutes under a cloth so the surface of the dough doesn't dry out.
2. Preparing the filling
   Mix the peanuts, broken pieces of walnuts, dried melon strips, dried fruit, cranberries, raisons, roasted flour, and sugar together. Pour the mixture on the honey and maltose and combine together to form a square.
3. Shaping the Shaobing
   Roll the skin dough into a long cylinder, begin to separate the dough with thumb and fore finger into sections of 25g each. Flatten one section into a round shape wrapper, now place a piece of the dry oil shortening (25g) into the wrapper. Wrap it well to form a ball. Repeat with process with the rest of the dough.
   Take the mixed fruit and nut filling that was prepared before and separate it into 25g pieces.
   Using a rolling pin, roll the dough ball from the top to the bottom, fold it three times, and roll it again to strengthen the dough. Repeat this process to strengthen the dough further then wrap around the fruit and nut fillings to form another ball.

Press down on the ball with your palm to make a disk that is about 2cm in height. Using a brush paint each disk with water and cover the surface with sesame seeds.
4. Cooking
Place each shaobing onto a tray ensuring enough space to cook evenly and bake at 180°C for 15 minutes, place on a plate and serve.

EGG YOLK SHORTCAKE

Ingredients

For the dry oil shortening dough
Cake flour 500g
Lard 250g

For the skin dough
Plain flour 500g
Lard 150g
Water 200g

For the filling
Preserved duck egg yolk 20
Red bean paste 300g

For brushing
Beaten egg yolks 50g

Method

1. Forming the dough
   To make the dry oil shortening, combine 250g of lard in 500g of cake flour together until a smooth dough has formed.
   To make the skin dough, mix 500g of cake flour, 150g of lard, and 200g of water together to form a dough. Leave it to rest for 15 minutes under a cloth so the surface of the dough does not dry out.
2. Shaping the shortcake
   Roll the skin dough into a long cylinder, begin to separate the dough with thumb and fore finger into sections of 25g each. Flatten one section into a round shape wrapper, now place a piece of the dry oil shortening (25g) into the wrapper. Wrap it well to form a ball. Repeat with process with the rest of the dough.
   Roll the red bean paste into a long cylinder and separate the dough into balls of 25g each. flatten the ball into a round shape wrapper. Place a duck egg in the centre of the wrapper and wrap it around to form a ball.
   Using a rolling pin, roll the dough ball from the top to the bottom, fold it three times, and re roll it to strengthen the dough. Repeat this process to strengthen the dough further then wrap around the red bean paste and duck yolk filling.
3. Cooking
   Once all the balls have been processed place them evenly onto a baking tray and brush them with egg wash. Bake at 180°c for 15 minutes, place on a plate and serve.

Chinese Culinary Techniques

BUDDHA HAND SHORTCAKE

Ingredients

For the dry oil shortening dough
Cake flour 500g
Lard 250g

For the skin dough
Cake flour 500g
Lard 150g
Water 200g

For the filling
Red bean paste 500g

For brushing
Beaten egg yolks 50g

Method

1. Forming the dough
   To make the dry oil shortening, combine 250g of lard in 500g of flour together until a smooth dough has formed.
   To make the skin dough, mix 500g of flour, 150g of lard, and 200g of water together to form a dough. Leave it to rest for 15 minutes under a cloth so the surface of the dough does not dry out.
2. Shaping the shortcake
   Roll the skin dough into a long cylinder, begin to separate the dough with thumb and fore finger into sections of 25g each. Flatten one section into a round shape wrapper, now place a piece of the dry oil shortening (25g) into the wrapper. Wrap it well to form a ball. Repeat with process with the rest of the dough.
   Roll the red bean paste into a long cylinder and make it into balls of 30g each.
   Using a rolling pin, roll the dough ball from the top to the bottom, fold it three times, and re roll it to strengthen the dough. Repeat this process to strengthen the dough further then wrap around the red bean paste and form a ball again. Flatten it to a disk that is about 2cm high and begin to score around the circumference of the disk. Score the dough in quarters making sure that each quarter has 4 scores and overall, 16.
   Take each ball and flatten on the angle leaving a thicker end on one side. The thick side should be about 1.5cm high and the thinner side is 0.5cm high. Score the thinner side like a clam shell towards the thicker side, score every 0.5cm to create a fan. Place a plastic dough scrapper into the centre of the fan to push it into the centre. Tuck any scored pieces of dough under the thicker part of the pas-

try.
3. Baking
Once all the balls have been processed place them evenly onto a baking tray and brush them with egg wash. Bake at 180°c for 15 minutes, place on a plate and serve.

YUAN BAO SHORTCAKE

Ingredients

For the dry oil shortening dough
Plain flour 500g
Lard 250g

For the skin dough
Plain flour 500g
Lard 150g

For the filling
Red bean paste 500g

For brushing
Beaten egg yolks 50g

Method

1. Making the shortcake dough
   To make the dry oil shortening, place 250g of lard in 500g of plain flour and begin to combine together with until a smooth dough has formed.
   To make the skin dough, combine 500g of flour, 150g of lard, and 200g of water together to form a dough. Leave it to rest for 15 minutes under a cloth so the surface of the dough does not dry out.

2. Shaping the shortcake
   Roll the skin dough into a long cylinder, begin to separate the dough with thumb and fore finger into sections of 25g each. Flatten one section into a round shape wrapper, now place a piece of the dry oil shortening (25g) into the wrapper. Wrap it well to form a ball. Repeat with process with the rest of the dough.
   Roll the red bean paste into a long cylinder and separate it into small pieces of 30g each.
   Using a rolling pin, roll the dough ball from the top to the bottom, fold it three times, and re roll it to strengthen the dough. Repeat this process to strengthen the dough further then wrap around the jujube paste and form a ball again. Flatten either side of the dough ball to create two flaps, fold each flap up towards the centre of the dough ball and correct the dough ball to form a neat shape.

3. Baking
   Place all the shortcakes on to a baking tray and egg wash the surface of the pastry and bake at 180°c for 15 minutes. Once cooked place on a plate and serve.

FLOWER SHORTCAKE WITH JUJUBE FILLING

Ingredients

For the dry oil shortening dough
Cake flour 500g
Lard 250g

For the skin dough
Cake flour 500g
Lard 150g
Water 200g

For the filling
Jujube paste 500g

For brushing
Beaten egg yolks 50g

Method

1. Forming the dough
   To make the dry oil shortening dough, combine 250g of lard in 500g of flour together until a smooth dough has formed.
   To make the skin dough, mix 500g of flour, 150g of lard, and 200g of water together to form a dough. Leave it to rest for 15 minutes under a cloth so the surface of the dough does not dry out.
2. Shaping the shortcake
   Roll the skin dough into a long cylinder, begin to separate the dough with thumb and fore finger into sections of 25g each. Flatten one section into a round shape wrapper, now place a piece of the dry oil shortening (25g) into the wrapper. Wrap it well to form a ball. Repeat with process with the rest of the dough.
   Roll the jujube paste into a long cylinder and separate it into small pieces of 30g each.
   Using a rolling pin, roll the dough ball from the top to the bottom, fold it three times, and re roll it to strengthen the dough. Repeat this process to strengthen the dough further then wrap around the red bean paste and form a ball again. Flatten it to a disk that is about 1.5cm high and begin to score around the circumference of the disk. Score the dough in quarters making sure that each quarter has 4 scores and overall, 16, open each score to empathise the flower shape.
3. Baking
   Place all the shortcakes on to a baking tray and egg wash the centre of the fanned pastry together and bake at 180°C for 15 minutes. Once cooked place on a plate and serve.

Chinese Culinary Techniques

SHUANGQIAO SHAOBING

Ingredients

For the dry oil shortening dough
Plain flour 400g
Lard 200

For the skin dough
Plain flour 400g
Yeast 4g
Water 240g

For the filling
Raw pork fat 200g
Roast flour (pre-roasted in the oven at 180°C until it turns golden) 200g
Sugar 250g
Roasted sesame seeds 200g
Raw sesame seeds 100g

Method

1. Forming the dough
   To make the dry oil shortening dough, combine 200g of lard in 400g of plain flour together until a smooth dough has formed.
   To make the skin dough, combine 400g of flour, 4g of yeast, and 240g of water together to form a dough. Leave it to rest for 15 minutes under a cloth so the surface of the dough does not dry out.

2. Preparing the filling
   Mix the raw pork fat and the sugar by rubbing them together with your fingertips, then add roast sesame seeds and rub again to combine.

3. Shaping the Shaobing
   Take the skin dough and open it out so you can place the dry oil shortening dough within it and form a ball. Now, begin to roll it out and fold the bottom into the centre and then the top into the centre. Roll the dough out into a long rectangle and then roll it from the top to the bottom lengthways. This is incorporating the doughs together and strengthening the dough.
   Separate the dough with your thumb and fore finger into sections of 25g each, roll them out, place the filling into the centre of the dough and form a ball.
   Roll the balls out to form an oblong that is around 1cm in thickness, brush with water and dip each oblong into the raw sesame seeds. Place them face down on to the worksurface to flatten the seeds evenly on the dough.

4. Cooking
   Place all the Shaobing on to a baking tray evenly to make sure they bake correctly. Bake at 180°C for 20 minutes, place on a plate and serve.

Part Three

THOUSAND-LAYERED PASTRY WITH CARAMEL TOPPING

Ingredients
Plain flour 500g
Water 250g
Eggs 5

Margarine 300g
Red bean paste 200g
Icing sugar 100g

Method

1. Forming the dough
   Place the flour on the worktable surface and make a hole in the middle with your hand, add the sugar, 1 egg and 250g of water into the hole. Slowly incorporate the surrounding flour with the water reaching a stage that is called "ears of the wheat" which is when the flour comes together with the water to form a breadcrumb stage. This is when you can knead until a dough is formed. Be careful not to add too much water as you will flood the dough. Continue to form the dough, adding more water if necessary, constantly feel the dough making sure that is not to dry or firm. Once the dough has formed continue to develop the gluten in the dough by kneading until the surface is smooth. Leave to rest for 15 minutes with a damp cloth over the top of the dough to prevent it from dry and forming a crust on the surface.

2. Laminating the dough
   Take the rested dough and roll it until a large rectangle, ensure the margarine is rolled out to 1cm thick and it is half the size of the pastry. Place on top of the rectangle of dough and enclose the margarine, make sure there is no doubled-up area of dough and that it is evenly enclosed.
   Roll the dough out to 1.5cm thick in a shape of a long rectangle. Fold the two ends into the centre and fold in half, this is known as a "book turn". Repeat this process and roll out gain to a large rectangle around 1.5cm in thickness and cut the rolled-out pastry into 7cm squares.
   If the pastry is becoming too warm or the margarine is coming through the dough place it in a fridge for 10 minutes to relax and firm. Make sure the margarine does not become too hard as it will not roll out evenly. Remember the margarine must be at a stage called "plastic". This is where the margarine is flexible but not too hard that it will snap if bent.

3. Forming the Thousand-layered Pastry with Caramel Topping
   Segment the bean paste into pieces that will fit into the laminated dough. Place the bean paste into the dough and old over dough to create a triangle. Place each triangle onto a baking tray with the correct space between each item.

4. Baking
    Brush egg wash onto each piece of dough and bake in an oven at 190°C and bake until golden brown and light to touch.
    For the decoration place 4 egg white in a mixer and whisk icing sugar in until a smooth pipeable paste is formed and pipe onto the baked items to serve.

Part Three

COCONUT TART

Ingredients

Plain flour 500g
Water 250g
Eggs 5
Margarine 450g
Caster sugar 250g
Desiccated coconut 150g
Milk powder 75g
Cheese powder 25g

Method

1. Forming the dough
   Place the flour on the worktable surface and make a hole in the middle with your hand, add 50g of the sugar, egg, gradually add 250g of water, slowly incorporate the surrounding flour with the water and the egg reaching a stage that is called "ears of the wheat" which is when the flour comes together with the water to form a breadcrumb stage. This is when you can knead until a dough is formed. Be careful not to add too much water as you will flood the dough. Continue to form the dough adding more water if necessary, constantly feel the dough making sure that is not to dry or firm. Once the dough has formed continue to develop the gluten in the dough by kneading until the surface is smooth. Leave to rest for 15 minutes with a damp cloth over the top of the dough to prevent it from dry and forming a crust on the surface.
2. Laminating the dough
   Take the rested dough and roll until a large rectangle, ensure the margarine is rolled out to 1cm thick and it is half the size of the pastry. Place on top of the rectangle of dough and enclose the margarine, make sure there is no doubled-up area of dough and that it is evenly enclosed.
   Roll the dough out to 1.5cm thick in a shape of a long rectangle. Fold the two ends into the centre and fold in half, which is known as a "book turn". Repeat this process and roll out again to a large rectangle around 0.75cm in thickness and cut the rolled-out pastry into 8cm circles using a cutter.
   If the pastry is becoming too warm or the margarine is coming through the dough place it in a fridge for 10 minutes to relax and firm. Make sure the margarine does not become too hard as it will not roll out evenly. Remember the margarine must be at a stage called "plastic". This is where the margarine is flexible but not too hard that it will snap if bent.
3. Preparing the filling
   Crack 3 eggs into a bowl and stir in the remaining 200g 0f sugar to combine, add the milk and cheese powder, desiccated coconut and the margarine and combine.
4. Forming the coconut tart and baking
5. Place a disk of pastry into a fluted tart ring and fill with the coconut filling and place onto a baking tray. Ensure the correct space is around each tart to cook evenly. Bake in an oven at 200°C for 20 minutes until golden brown.

RICE ROLLS WITH SESAME SEEDS

Ingredients
Sticky rice 500g
Lotus seed paste 300g
Sesame seeds 250g

Method
1. Preparing and cooking the rice
   Wash the rice and soak it in cold water for 30 minutes. In another serving bowl, totally cover the rice with water. Place in the oven to steam for 30 minutes until fully cooked. Tip the cooked rice out of the bowl onto a clean cloth and work the rice until it is malleable and forms a dough-like appearance.
2. Plating the dessert
   Toast the sesame seeds and begin to fold them into the cooked rice by rolling them into the dough. Form the lotus seed paste into a rectangle rolled out in clingfilm and place it on top of the sesame seed rice, fold over the top edge and roll to the centre. Repeat this with the bottom by rolling it to the centre. Turn the rolled rice over so the folds are on the base and flatten the top side and ensure that the roll is even. Slice every 3cm wide and place on a plate to serve.

EIGHT-TREASURE RICE PUDDING

Ingredients
Glutinous rice 100g
Lotus seed paste 500g
Sugar-marinated dried fruit 200g
Sugar-marinated Chinese goji berries 75g
Sugar-marinated osmanthus 20g
Caster sugar 800g
Lard 200g
Corn flour water 50g

Method

1. Preparing and cooking the rice
   Wash the rice and soak in cold water for 30 minutes. In a suitable seized serving bowl cover in lard and decorate in an attractive pattern with the Sugar-marinated dried fruit and goji berries.
   In another serving bowl fill with rice and water above the rice. Place in the oven for and steam for 30 minutes until fully cooked. Tip out of the bowl on to a clean cloth and work the rice until it is malleable and forms a dough like appearance.

2. Plating the dessert
   Form a ball with the rice and place the lotus seed paste in the middle and form the rice around the Lotus seed paste. Now place the rice ball in the pre-prepared serving bowl that has been decorated, push the rice down and over all the decorated areas with the rice ball. Place back into the oven set on steam and cook for 20 minutes, then turn out onto a serving plate.
   Begin to make a sugar solution with a ladle of water in a wok, add sugar and sugar-marinated osmanthus and begin to reduce the water and sugar solution and add the corn flour water and cook out for 3 minutes. To serve ladle over the steam rice.

GLUTINOUS RICE BALL WITH COCONUT SPRINKLES

Ingredients

Glutinous rice flour 500g
Wheat starch 150ml
Lotus seed paste 300g
Caster sugar 200g
Lard 15g
Coconut sprinkles 100g
Cranberries 20g

Method

1. Making the Filling
   Place the rice flour and wheat starch into a bowl, add the sugar and lard and mix adding water when needed to form a thick liquid. Steam for 30 minutes and remove from the oven and pour on top of the desiccated coconut and cover entirely with the coconut.
   Start to divide the dough into 30g pieces and ensure that they are covered in the desiccated coconut.
2. Forming the rice ball
   Split the lotus seed paste into 30g pieces and place a piece into the centre of each rice flour dough. Form into a ball and roll again in the desiccated coconut. To finish garnish with a small piece of cranberry and serve.

NINGBO STYLE GLUTINOUS RICE BALL

Ingredients

Glutinous rice flour 500g
Caster sugar 150g
Pork fat 150g
Black sesame seeds 150g
Water 280g
Walnuts 150g

Methods

1. Preparing the ingredients
   Crush the sesame seeds with a rolling pin, finely chop the walnut.
2. Making the filling
   Mix the sugar, pork fat, sesame seeds, walnuts, using your fingertips to combine them to form a paste. Roll the paste into a rectangle and cut it into cubes that are around 2cm. Leave them aside for the next stage.
3. Forming the dough
   In a mixing bowl, mix the rice flour with water to make the glutinous rice dough. Take enough of the dough and wrap it around a small piece of the sesame seed paste to form a ball.
4. Cooking
   Once all the balls are formed, cook them in boiling water on high heat for 3 minutes. Add in some cold water and continue to cook on high heat for 3 minutes. Serve with soup.

Chinese Culinary Techniques

CHERRY BLOSSOM

Full ingredients list
Double cream 358ml
Caster sugar 71g
Egg yolk 95g
Vanilla pods 1
Cherry flavouring 31.5g
Agar agar 1g
Cherry Puree 142ml
Caster sugar 1.5g
Fresh cherries 28g
Moltosec 100g
White chocolate 75g
Salt 1g
Double cream 225ml
Yolks 90g
Caster sugar 22g
Ruby Chocolate 337g
Whipping cream 450ml
Red colouring 0.5g

CHERRY CRÈME BRÛLÉE INSERT

Ingredients
Double cream 358ml
Caster sugar 71g
Egg yolk 95g
Vanilla pods 1

Method
Whisk the sugar and egg yolks together until the sugar is complete dissolved within the egg yolks and leave to one side. Boil the double cream with the vanilla beans so that the vanilla is infused into the cream. Pour half the boiled double cream over the sugar and yolks and incorporate together, return the cream and egg yolks to the heat and constantly stir until it reaches 82°C. Pour into a suitably seized mould and bake in the oven at 100°C until there is a slight wobble in the centre of the brûlée. Freeze until solid and remove from the mould to use once the ruby chocolate soil is ready.

CHERRY COMPOTE

Ingredients
Agar agar 1g
Cherry Puree 142ml
Caster sugar 1.5g
Fresh cherries 28g

Method
Add the agar agar and sugar in a bowl and mix together, in a saucepan boil the cherry puree and pour half the hot puree over the sugar and agar agar mixture and incorporate it together. Return the cherry puree into saucepan with the remaining puree and bring it back to a boil. Once cooled, blend in a food processor until smooth

Part Three

and then add the fresh cherries once chopped.

RUBY CHOCOLATE SOIL
Ingredients
Moltosec 100g
White chocolate 75g

Method
Melt the chocolate and add the moltosec (tapioca maltodextrin is derived from tapioca starch). Mix until a dry powder is formed, if it is too wet add more moltosec until it is dry. Continue to rub between your fingers until the powder is fine and add the salt to season.

Chinese Culinary Techniques

PEACH OF IMMORTALITY

Full ingredients list
Diced nectarine 150g
Peach for compote 150g
Ascorbic acid 0.6g
Lime juice 6ml
Caster sugar 6g
Pectin 2.4g
Coconut puree 237g
Caster sugar 45g
Gelatine 8g
Double cream 338g
Moltosec 100g
White chocolate 75g
Salt 1g
White chocolate 33g
Cocoa butter 50ml
Red or yellow food colour 21g

NECTARINE INSERT
Ingredients
Diced nectarine 150g
Peach for compote 150g
Ascorbic acid 0.6g
Lime juice 6ml
Caster sugar 6g
Pectin 2.4g

Method
Begin by peeling and fine dicing one half of the nectarines and cover in the lime juice and ascorbic acid to prevent them from browning. Leave covered with clingfilm touching the diced nectarines until required. Peel and destone the other half of the nectarines, and dice.
Place into a saucepan until the flesh of the nectarine starts to break down into a puree consistency. Then add the puree in the food processor and blend until the puree is smooth. Take the puree and heat once again, then pour 500ml of the puree over the pectin and sugar and incorporate it until smooth. Return it to the rest of the puree and bring it to the boil. Fold both the the nectarine products together and pour into the correct moulds and freeze. Once frozen, demould and leave in the freezer until the required.

COCONUT MOUSSE
Ingredients
Coconut puree 237g
Caster sugar 45g
Gelatine 8g
Double cream 338g

Method
To begin, whip the coconut cream to soft peaks and leave in the fridge until needed. Hydrate the gelatine in a

saucepan in cold water, once soft remove from the water and strain the excess water. In a saucepan boil the coconut puree and sugar and add the gelatine and incorporate it until. dissolved. Ensure the coconut mixture has cooled to 30°C and fold in the pre-whipped cream into the coconut puree gently. Ensure that all the coconut puree is mixed with the cream.

WHITE CHOCOLATE SOIL
Ingredients
Moltosec 100g
White chocolate 75g
White colouring 5g
Salt 1g

Method
Melt the chocolate and add the moltose (tapioca maltodextrin is derived from tapioca starch) with the white colouring. This must be a fat-soluble colouring. Mix until a dry powder is formed, if it is too wet add more moltosec until it is dry. Continue to rub between your fingers until the powder is fine and add the salt to season.

COLOURED COCOA SPRAY
Ingredients
White chocolate 33g
Cocoa butter 50ml
Red or yellow food colour 21g

Method
If you have a thermomixer place all the ingredients into the thermo, mix on speed 3 for 10 minutes at 80°C. If the thermomixer is not available, place all the ingredients into a bain marie and mix until all ingredients have melted and fully incorporated. Sieve and leave to one side prior to use. Repeat this process with both colours.

ASSEMBLY
Pipe the coconut mousse halfway up the peach mould, insert the frozen nectarine insert and cover completely with the coconut mousse. Smooth the off the tops of the mould and freeze, once frozen demould and spray with a spray gun the peach. The peach must be frozen so when the cocoa spray is blown onto the peach it instantly freezes creating a powder coat. Spray with a base coat of yellow and add the red colouring to create the appearance of a peach.
To place the white chocolate soil onto a plate and place the peach onto the centre, Place two verbena leaves into the centre of the peach.

Section 7 Plate Presentation Ideas

Corner Style is a decorative technique that places the main ingredients on one side or at one corner of the plate. Such a method is suitable for decorating large or whole sized food such as Eight-treasure duck, Roast lamb leg, Braised mandarin dish. Food is normally served on round, oval, or rectangle plates. Sample designs are

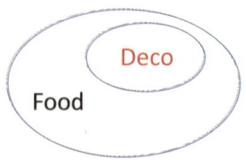

Mountain Shape
Ingredients: baby cucumber 1, sausage 1/2, boiled carrot 1/2, dragon fruit 1/6, fresh lemon 1/3, cherry 1, parsley 1, mash potato 100g.

Floral Hoops
Ingredients: celery stem 1, boiled carrot 1, red pepper 1/6, sunflower (or butterfly orchid) 1, golden cane palm leaf 1, mash potato 100g.

Asymmetrical Style is also known as three-point or five-point decoration. It is frequently seen in dry dishes with shredded, diced, sliced, stripped or score-cut food.

Sample designs are
Three Flowers and Leaves
Ingredients: baby cucumber 1, cherries 3.

Five Lotus Flowers
Ingredients: baby cucumber 1, boiled carrot 1, yellow cherry tomatoes 2.

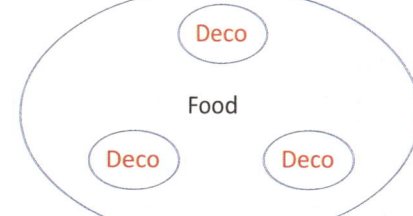

Symmetrical Style where the plate is decorated with symmetrical and identical patterns. There are the single-paired style (A), two-paired style (B), and multi-paired style (C).

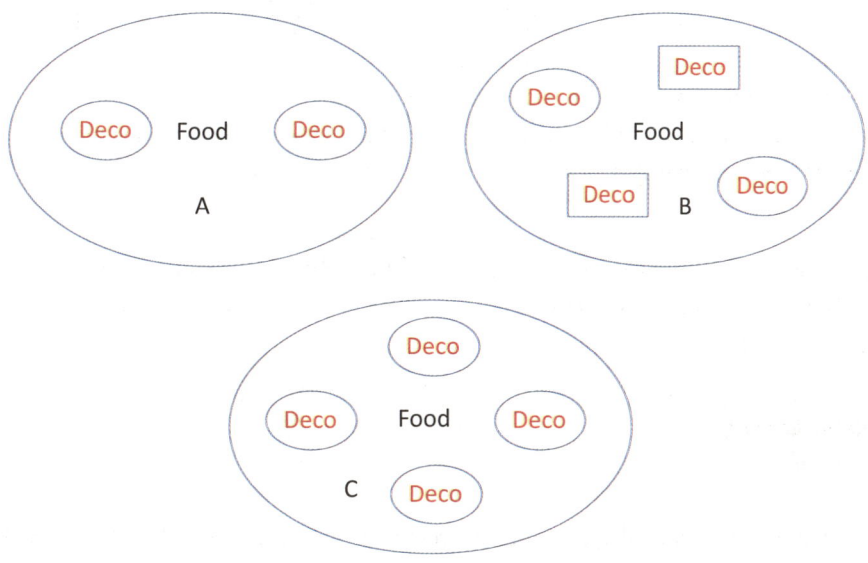

Sample designs include

Peach Hearts
Ingredients: red delicious 1, baby cucumber 1/2, dragon fruit 1/6, mash potato 50g.

Fruit Salad
Ingredients: mash potato 100g, golden cane palm leaf 1, Chinese cypress sprig 1, cherries 2, dragon fruit 1/6.

Centre Piece Style. In this style, a three-dimensional piece is often placed at the centre of the plate to highlight the theme of the dish and the cooked ingredients are arranged around it. For example, a dish called "Yipin Vegetarian Stew" in which a variety of rare mushrooms are used. In order to highlight the vegetarian nature of this dish, the centre piece is an arhat figure with crossed legs carved from carrots. Sample designs include

Part Three

Flower Blossoms
Ingredients: carnation 1, baby's-breath 2, golden cane palm leaf 1, purple perilla leaves 2, Chinese cypress sprig 1, mashed potato 50g, green/ yellow/ purple pectin (10g each).

Dancing Butterfly
Ingredients: pumpkin 1 piece, boiled carrot 1/2, radish 1/8, white radish (Mooli) 1 piece, parsley 1.

Division Style in which the decorative ingredients are used to form up a belt to divide the plate into two halves. Two different flavoured dishes can be served on one plate. For example,

Flower Swirl
Ingredients: Xinlimei radish 1, baby cucumber 1, mash potato 50g.

Half Rim Style where decorative ingredients are laid out in patterns around the rim of the plate about halfway through. The cooked ingredients take up 2/3 of the plate. For example, Fruitful Tree
Ingredients: winter melon (wax gourd) skin 5 pieces, cherry 10, garlic sprouts 2.

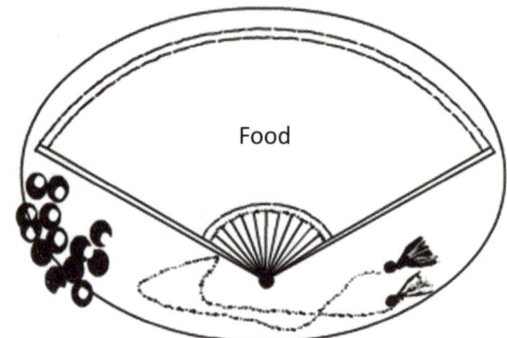

Pictographic Frame Style. In this style, ingredients are cut and arranged to outline pictographic patterns such as peaches, fans, flower baskets, lanterns, and fish. Cooked ingredients then go on to fill up the space. This type of design is often based on the natural colours and shapes of the decorative ingredients. Sample designs include

Fan
Ingredients: 1 large piece of winter melon (wax gourd) skin, cooked carrots 1/2, baby cucumber 1, fried beaten egg 1/4 piece.

Longevity Peach
Ingredients: cooked carrot 1, baby cucumber 1, blanched rape green leaves 2.

Fruit and Vegetable Carving
Underwater Landscape
Ingredients
Boiled carrot 1/2, white radish (mooli) 1/2, pumpkin 1 rectangle piece, winter melon skin 1, parsley 1.
Tools
Cleaver, carving knife, small and medium-sized U-shape poker, toothpick.
Process
1. In the middle to the left side of the ingredient, carve two arcs upwards and downwards, and make the edge thinner.
2. Carve a shape of the fish lips, and smooth it.
3. Carve two fish barbels and the bodies.
4. Carve the fish heads and gills, and smooth it.
5. Carve the fish fins.
6. Carve tail fins.
7. Carve the contours of the fish and the lines in the fins (keep smoothing the carves).
8. Carve the fish scales (keep smoothing the carves).

9. Carve the patterns on the tails (fish carving completed).
10. Create the coral with a U-shaped poker and insert it on the rectangular pumpkin base with toothpicks. Create some sea weeds with the winter melon skin and insert them under the coral. And insert the fish on the coral and garnish it with sprigs of parsley.

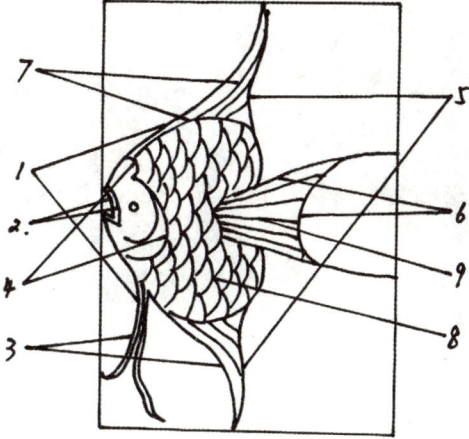

Flower

Ingredients

Xinlimei radish 1/2, white radish 1 triangle piece, winter melon skin 3 pieces.

Tools

Cleaver, carving knife, medium-sized U-shape poker, toothpick.

Process

1. Rinse the radish and remove its two ends, halve it horizontally.
2. Decapitate the radish and put it downward and cut a pentagon.
3. Create arc-shaped petals based on the pentagon with a carving knife.
4. Remove excess waste between the two petals.
5. Carve more petals between the two petals.

6. Repeat the 5th step and create more layers of petal (stop carving until the carving knife should erect vertically when used).
7. Carving the stamen is the same as carving the petals, but the stamen is smaller so petals should be less (3 pieces are perfect). Remove the extra part in the middle of the stamen to make a blooming flower.
8. When the carving is finished, soak the flower in alum water and a peony is done.
9. Make some small mountains and rocks out of the white radish with the medium-sized U-shaped poker, and then fix the peony flower with a toothpick on the rocks, and finally make some leaves out of the wax melon skins and tie them around the bottom of the peony.

Part Three

Chapter 11 Dough Modelling

Section 1 An introduction to Tianjin Style Dough Modelling

Dough modelling, or dough figurine, is a wonder among Chinese folk arts using simple tools and materials such as wheat flour and glutinous rice flour to create colourful lifelike figurines.

The earliest form of dough figurine work is thought to be the "folk dough flower", a type of flower-shaped steam bun that was created in the Han Dynasty. This indicates that people began to generate aesthetic ideas in presenting food hence food was no longer merely to fill the stomach.

Originated from the Heze dough modelling in the Qing Dynasty, Tianjin dough modelling developed its unique artistic features in 1947 during the period of Republic of China (1912-1949) and firmly established its distinctive style in 1985. Since the end of the 1990s, Tianjin dough modelling has become an emerging style in the dough modelling society with its detail-oriented and complex modelling and shaping techniques, simply and traditional design styles, elegant and natural colour combination.

The materials and tools required to make dough figurines are remarkably simple. Using techniques such as kneading, rubbing, pinching, and shaping with hands, dough modelling artists can create a wide variety of subjects in the forms of human figures, animals, flowers, feathers, fruits, and vegetables.

Through these figurines, artists express their spiritual world and good wishes for life. The themes of dough modelling artworks are often based on Chinese folk myths and legends, historical stories, Buddhism and Taoism teachings, drama and novels, and classic literatures such as *Uproar in Heaven, Liu Hai Plays with the Golden Toad, Kylin Songzi, The Eight Immortals Crossing the Sea, Legend of the White Snake, The Drunken Concubine, Farewell to My Concubine, Water Margin, Romance of the Three Kingdoms* etc. These dough modelling artworks provide with people not only a unique art form to appreciate but also a means to keep the traditional folk stories alive. As a result, dough modelling is regarded as an intangible cultural heritage.

Tianjin dough modelling is an important member of Chinese dough modelling society with the following characteristics.

1. Use a wide range of natural colour shades to create simple, smooth, coherent, and balanced colour themes.
2. Artworks are created in various shapes, sizes, and story themes. In terms of the shape types, there are the stick figurines
   a. Flat figurines.

   b. Stick figurines are the traditional form. Created on a bamboo stick, this type of figurines is small in size and simple in design hence it's ideal for carrying around.

   c. Framed figurines. This type of dough modelling artworks is created on top of a frame that are made of thin bamboo stripes or metal wires. They are larger in size with a bigger visual impact.

   d. Integral figurines. This type of artworks is entirely made of dough. Smaller in size comparing with

Part Three

the framed figurines, it embodies more profound cultural expression of Chinese folktales.

3. Good at creating a vivid facial structure and expression. Traditionally, impressionism dominates the dough figurines artworks. Tianjin dough modelling, however, utilises some realism techniques of western sculpture to rejuvenise this traditional art form. Their artworks have the characteristics of both impressionism and realism in sculpture.
4. Tianjin dough figurines emphasis on the accuracy and fluency of cloth and accessories, bones, and muscles to express emotions of the characters created.
5. Good at forming scenes by creating appropriate props, group characters with appropriate facial expressions.
6. The accessories are made of different materials and in a range of patterns. They are created based on the story themes.

Section 2 Materials & Tools for Dough Modelling

The dough for Tianjin dough modelling is made of wheat flour, glutinous rice flour, preservatives, desiccant, and antioxidant. The high-quality wheat and rice selected to make the dough are produced locally in Tianjin. The smooth and flexible semi-transparent dough is a highly plastic material. It sustains its colour very well once is dyed. Therefore, it is ideal to create lifelike skin texture, cloth that looks natural, beautiful silky hair, and flexible belts.

List of ingredients for the dough.

Flour: all-purpose wheat flour, which is low in elasticity, high in plasticity, and durable once flour dough is formed.

Glutinous rice flour: high stickiness to prevent the artworks from cracking.

Potassium sorbate: edible preservative

Calcium propionate: edible preservative

Sodium benzoate: edible preservative

Sorbitol: desiccant

Glycerol: desiccant

Corn syrup: desiccant

Dough modelling artists tend to design their own tools which include sculpture knife, eye opener, pressing plate, poking knife (various in materials such as metal, ox bone, organic glass, plastic, etc.), scissors, tweezers, comb, fountain pens, colouring pens. Sculpture knife is applied for modelling, carving faces, shaping, and carving clothing patterns.

Part Three

- Eye opener is used to open eyes, the mouth or separate tiny thin dough pieces.
- Poking knife is used to create patterns on the armor, fish scales, or other line patterns.
- Pressing plate is used to press the dough into different sizes of sheets.
- Scissors are used to cut open the dough.
- Comb is used to make lines or beaded laces.
- Tweezers and colouring pens are applied for final decoration.
- Fountain pen is applied to smooth, moisturize and clean the surface.
- The white glue is applied to connect any two parts so that the works will keep its entirety after being dried.

Section 3 Fundamental Dough Figurines Making Techniques

Widely used fundamental carving and shaping methods include kneading, pinching, rolling, pressing, rubbing, poking, lifting, grinding, pushing, and carving

3.1 Carving and Shaping Methods

Method	Description
Kneading	Use both hands to massage and combine a piece of dough from surrounding area to the centre repeatedly until its surface is smooth and plump.

Chinese Culinary Techniques

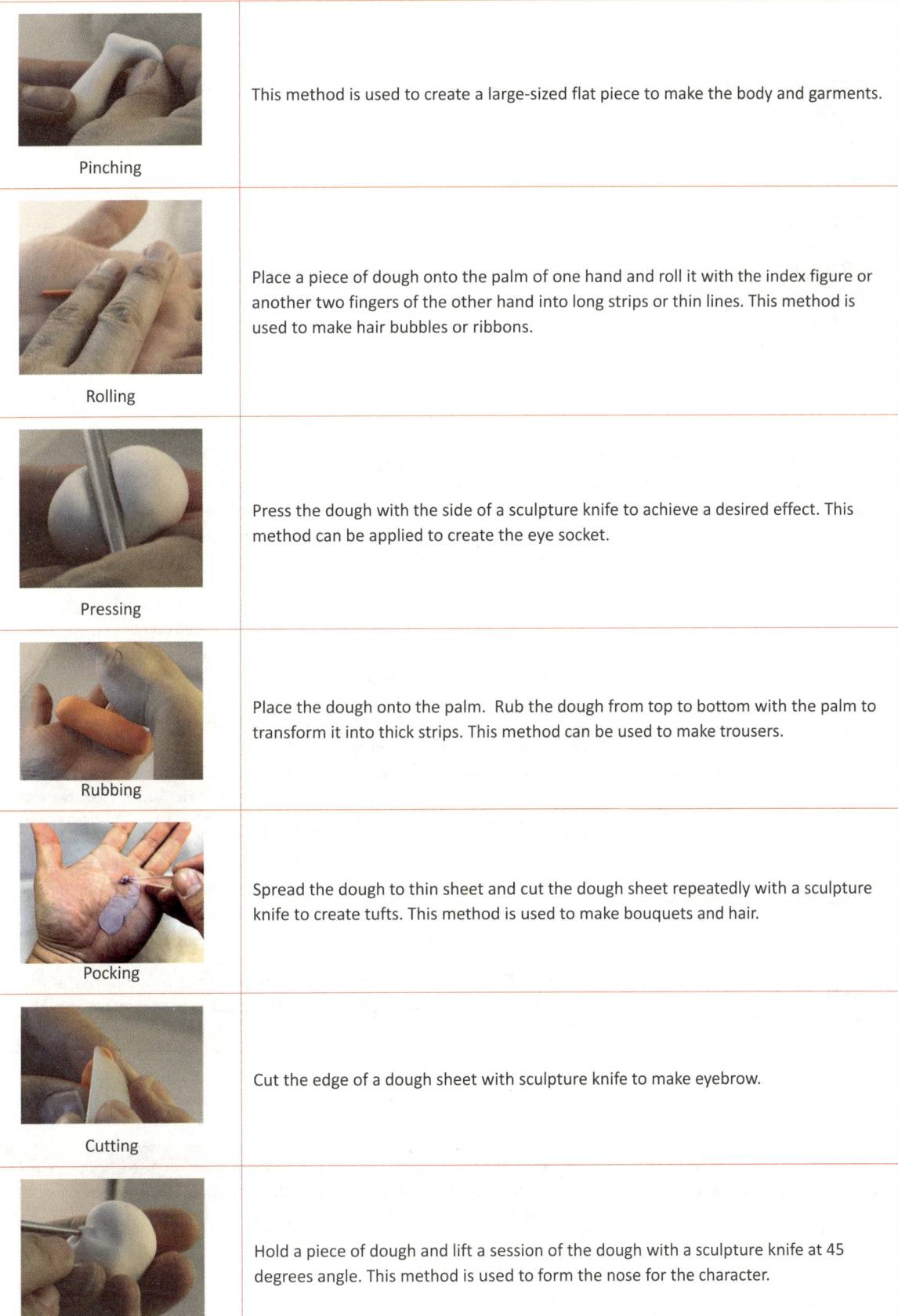

Pinching	This method is used to create a large-sized flat piece to make the body and garments.
Rolling	Place a piece of dough onto the palm of one hand and roll it with the index figure or another two fingers of the other hand into long strips or thin lines. This method is used to make hair bubbles or ribbons.
Pressing	Press the dough with the side of a sculpture knife to achieve a desired effect. This method can be applied to create the eye socket.
Rubbing	Place the dough onto the palm. Rub the dough from top to bottom with the palm to transform it into thick strips. This method can be used to make trousers.
Pocking	Spread the dough to thin sheet and cut the dough sheet repeatedly with a sculpture knife to create tufts. This method is used to make bouquets and hair.
Cutting	Cut the edge of a dough sheet with sculpture knife to make eyebrow.
Lifting	Hold a piece of dough and lift a session of the dough with a sculpture knife at 45 degrees angle. This method is used to form the nose for the character.

Part Three

 Grinding	Transform a piece of dough into sheets with the sculpture knife or pressing plate. This method can be used to make big dough sheet.
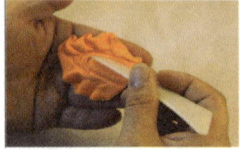 carving	Create lines and shapes with the pressing and pushing of sculpture knife onto the dough. This method is used to make textured clothing.

3.2 Fundamental Modelling Methods

Method	Description
Water drop making method	Rub a piece of dough into a ball, hold the dough between the fingers and sharpen one end to make it water drop-shaped. This method is used to make animal body or other parts
Marble making method	Mix pieces of dough of different colours to a multicoloured dough, separate the dough into pieces randomly and then combine them back together to make a dough ball or press it into a piece of dough sheet. This method can be used to make stones
Assorted tripes making method	Rub dough stripes of different colors (normally more than three) with hands to make its center thick and two ends pointed, put them together parallel into the left palm, join the palms together to rub the dough repeatedly until the flower strip is formed. This method is used to make lower hem or tail fin of a gown
Gradient colour making method	Place dough pieces of different colours together in sequence, flatten them with fingers into a piece of sheet, roll the sheet up from bottom to top, flatten it again. Repeat this process until the dough achieves a naturally smudged gradient colour. This method can be used to make juicy peaches
Ball making method to create animal figures	Divide a piece of dough into balls of different sizes according to the proportion of the head and body. This method is used to make animal figures
Egg-shaped making method to create the main body of animal figures	Form a piece of dough into a cylinder shape first and pinch it into oval shape, then press it with fingers to create the basic shape of an animal's main body
Glue method to assemble parts	Apply white glue onto the parts, joint them together, use a sculpture knife to roll and press on each side to tighten the joint. This method is used to assemble arms and legs

Chinese Culinary Techniques

Conversion tables (approximately)

WEIGHTS

Imperial (ounces)	Metric (grams)
¼ oz	7 g
1/2 oz	15 g
3/4 oz	21 g
1 oz	28 g
1 ¼ oz	35 g
1 1/2 oz	42.5 g
1 2/3 oz	45 g
2 oz	57 g
3 oz	85 g
4 oz (1/4 pound)	113 g
5 oz	142 g
6 oz	170 g
7 oz	198 g
8 oz (1/2 pound)	227 g
12 oz (3/4 pound)	340 g
16 oz (1 pound)	454 g
32.5 oz (2.2 pounds)	1000 g (1 kilogram)

VOLUMES

Imperial	Metric
1 fl oz	25 ml
2 fl oz	50 ml
3 fl oz	85 ml
5 fl oz	150 ml
10 fl oz	300 ml
15 fl oz	450 ml
1 pint	600 ml
1 ¼ pint	700 ml
1 ½ pints	900 ml
1 ¾ pints	1 Litre
2 pints	1.2 Litres
2 ¼ pints	1.25 Litres
2 ½ pints	1.5 Litres
2 ¾ pints	1.6 Litres
3 pints	1.75 Litres
3 ¼ pints	1.8 Litres
3 ½ pints	2 Litres
3 ¾ pints	2.1 Litres
4 Pints	2.25 Litres
5 pints	2.75 Litres
6 pints	3.4 Litres

Part Three

7 pints	3.9 Litres
8 pints (1 gallon)	5 Litres

OVEN TEMPERATURES

Degrees Fahrenheit	Degrees Celsius
200 °F	100 °C
250 °F	120 °C
275 °F	140 °C
300 °F	150 °C
325 °F	160 °C
350 °F	180 °C
375 °F	190 °C
400 °F	200 °C
425 °F	220 °C
450 °F	230 °C
475 °F	246 °C